ARAMCO, THE UNITED STATES, AND SAUDI ARABIA

ARAMCO
THE UNITED STATES
AND SAUDI ARABIA

A Study of the Dynamics
of Foreign Oil Policy
1933-1950

IRVINE H. ANDERSON

PRINCETON UNIVERSITY PRESS   PRINCETON, NEW JERSEY

To Donna and Mary

CONTENTS

PREFACE

ON MAY 29, 1933, Lloyd N. Hamilton of the Standard Oil Company of California and 'Abd Allah Al Sulaiman Al Hamdan, finance minister of Saudi Arabia, signed an agreement giving Socal exclusive rights to extract petroleum from an area larger than the state of Texas, in return for a small cash payment and royalties on future production of 4 shillings per ton (about $.10 a barrel in 1933). The agreement went virtually unnoticed in the rest of the world because almost no one—including experts from the other major oil companies and King 'Abd al-'Aziz ibn 'Abd al-Rahman Al Faisal Al Sa'ud himself—believed that there was oil in the newly formed kingdom of Saudi Arabia. The United States did not even have a consul—much less a diplomatic representative—in the country at the time. Only a handful of Socal geologists were convinced that oil was there, but events proved them to be right a thousand times over.

In 1936 Socal formed a partnership with The Texas Company to market Saudi oil, and responsibility for production was given to a joint subsidiary—later renamed the Arabian American Oil Company, Aramco. By 1941 test wells had begun to reveal the extent of reserves that had been tapped into, but at that point the American government still had little interest. For years the United States had been a net exporter of petroleum, and to many people this appeared to be the natural order of things.

In the midst of World War II, however, government planners came face to face with stark predictions that the United States would become a net importer after the war, and they were jolted into attempts to assure continuing American access to Saudi oil and rapid development of Saudi reserves to help conserve those in the Western Hemisphere. Through a Petroleum Reserves Corporation (PRC) created expressly for the purpose, Secretary of the Interior Harold L. Ickes tried to buy a controlling interest in the

Arabian American Oil Company. When that move failed, the
Department of State tried to accomplish the same objectives with
a negotiated Anglo-American Petroleum Agreement, but it failed
to win Senate ratification. Both initiatives ran aground on opposition from other American companies—chiefly the Texas independents—who feared competition from cheap Saudi oil. At the
end of the war, Ickes left the government, and the State Department turned to benign support of private enterprise to achieve its
objectives.

By 1946, Socal and Texas had decided that they lacked the
capital and the markets to develop the concession as rapidly as
'Abd al-'Aziz wanted it done, and they sold a minority interest
in Aramco to Standard Oil (New Jersey) and Socony-Vacuum.
The new partnership found itself immediately embroiled in the
problem of building a trans-Arabian pipeline to the Mediterranean
through territory in turmoil over the first Arab-Israeli War. By
this time, however, the United States was deep into the Cold War,
and the Department of State and National Security Council considered the economic welfare of Saudi Arabia a critical link in
the defensive arc around the Soviet Union. A de facto coalition
of government agencies, oil companies, and 'Abd al-'Aziz now
formed around the issue of developing Saudi reserves and its first
major accomplishment was modification of the 1933 Concession
Agreement in late 1950 to permit fifty-fifty profit sharing between
Aramco and the government of Saudi Arabia. By 1950 the style
of American Middle Eastern oil policy and what was later termed
the special relationship with Saudi Arabia were firmly established.
The seventeen years between 1933 and 1950 had been eventful
ones.

This book is an attempt to carefully reconstruct those seventeen
years, as a case study in the dynamics of American foreign oil
policy. In addition to material used by numerous previous writers
on the subject, it makes use of a considerable volume of corporate
documents acquired by the Federal Trade Commission in the early
1950s and by the Church Committee in 1974, as well as interviews
with thirty-three former corporate and government officials who

participated directly in the events described. A complete list is given in the Essay on Sources. The intent has been to produce a balanced account from both the government and the corporate side, as free from factual error as possible.

The book is written as a jargon-free narrative, but it is based on a method of social system analysis that focuses on the belief system, relative power, and specific objectives of each major participant at each decision point. Readers interested in such matters are referred to Appendix A, where the method is described in some detail. The systematic examination of a large body of data from this viewpoint called attention to a number of details and connections that might otherwise have gone unnoticed, and the resulting narrative is much the richer for it.

The reader will find no saints, no sinners, and no dark conspiracies in this account. Instead, he will find a long series of confrontations between competing interest groups within and without the American government—a process that led to the de facto coalition dedicated to increasing Saudi production. Whether or not this style of decision making and this type of coalition were in the real national interest may be a matter for debate, but whatever one's views on that subject, the process itself appears to have been typically American.

*

THIS STUDY was made possible by generous grants from the University of Cincinnati, the American Council of Learned Societies, and the National Endowment for the Humanities. So many people have contributed along the way that it is impossible to acknowledge my debt to all of them. I am especially grateful, however, to George Rentz of the Johns Hopkins School for Advanced International Studies and James V. Knight of Aramco's Washington office, who were most helpful in suggesting avenues to explore and people to contact at the very beginning of this study in 1976. My special thanks go to all of those who graciously agreed to interviews, and whose remembrances have greatly enriched this work. And I have a lasting debt to Bennett H. Wall of Tulane University for sharing with me his exploration of Exxon corporate records.

As the study developed, additional insight and suggestions came from Michael M. Ameen, Jr., of Mobil's London office, John A. DeNovo, of the University of Wisconsin, Wanda M. Jablonski, of the *Petroleum Intelligence Weekly*, Joseph J. Malone, of Middle East Research Associates in Washington, A. J. Meyer and Thomas R. Stauffer, of Harvard's Center for Middle Eastern Studies, Aaron D. Miller, of the Department of State, Elizabeth Monroe, in London, Edith Penrose, of the Institut Européen d'Administration des Affaires, Georgiana Stevens, in San Francisco, and Mira Wilkins, of Florida International University. George T. Ballou was especially helpful for background on Standard Oil of California, and various corporate statistics and publications were provided by A. C. DeCrane, Jr., of Texaco, Inc., W. K. Morris, of Standard Oil of California, I.P.H. Skeet, of Royal Dutch-Shell, and C. R. Williams, of Mobil. Joel Darmstadter, of Resources for the Future, was kind enough to share with me some difficult-to-find statistical data on world petroleum production, and Ivy Oneita Duce, of Walnut Creek, California, provided me with considerable background on her husband, James Terry Duce. I am also indebted to Eileen Bagus, Merton S. Krause, James A. Stever, and especially Alfred Kuhn for having read and critiqued an early version of the Methodological Note.

Professors Mira Wilkins, Joan Hoff Wilson, Gerald D. Nash, and George Rentz read and commented on the original manuscript. Their suggestions were most helpful, but I, of course, assume full responsibility for any remaining errors of fact or interpretation.

Support for this project took tangible form at the University of Cincinnati in the moral support of Dean Ernest G. Muntz of Raymond Walters College, the excellent bibliographic research of Patricia Mooney-Melvin in the early stages of the project, the skill and tenacity of Harriette L. Wolffe in locating obscure publications and old government documents, the expert indexing of Judith Spraul-Schmidt, and the indefatigable and always cheerful typing of Janet Winter.

Most important, my deep and lasting thanks go to my wife

Donna and my step-daughter Mary, without whose infinite patience, warm encouragement, and constant love this book would not have been possible.[1]

Irvine H. Anderson
Cincinnati, Ohio
July 1980

[1] Donna suggested addition of the term ''silent and long suffering'' at this point, but the length of time spent discussing the accuracy of the term ''silent'' finally precluded its use.

LIST OF TABLES

ABBREVIATIONS

Aminol	American Independent Oil Company
ANPB	Army-Navy Petroleum Board
API	American Petroleum Institute
Aramco	Arabian American Oil Company
Bapco	Bahrain Petroleum Company
Caltex	California Texas Oil Company
Casoc	California Arabian Standard Oil Company
C.F.P.	Compagnie Française des Pétroles
ECA	Economic Cooperation Administration
f.o.b.	free on board (that is, not including freight or insurance charges)
GATT	General Agreement on Tariffs and Trade
IPAA	Independent Petroleum Association of America
IPC	Iraq Petroleum Company
NEA	Office of Near Eastern and African Affairs
PAW	Petroleum Administration for War
PIWC	Petroleum Industry War Council
PRC	Petroleum Reserves Corporation
Socal	Standard Oil Company of California
SWNCC	State-War-Navy Coordinating Committee
Tapline	Trans-Arabian Pipeline Company
TPC	Turkish Petroleum Company

ARAMCO, THE UNITED STATES, AND SAUDI ARABIA

I

IN THE MIDDLE EASTERN LABYRINTH

SAUDI ARABIA first entered the American consciousness as a place of strategic importance in 1943, in the midst of World War II. As the scope and complexity of modern warfare gradually became apparent, planners in the Departments of the Navy, State, and the Interior took increasing note of two developments. First, a mid-war dip in the rate of annual domestic discoveries of oil below the rate of annual domestic production cast serious doubt on the sufficiency of American reserves in the years ahead.[1] And second, reports began to circulate of vast untapped reserves in Saudi Arabia in a concession held by two American companies—Standard Oil of California and The Texas Company. The logical consequence was quickened interest in a firm, long-term connection between the United States and the oil of Saudi Arabia. The oil, however, was presided over by an astute and most unusual monarch, 'Abd al-'Aziz ibn 'Abd al-Rahman Al Faisal Al Sa'ud,[2] whose desert kingdom was unlike anything in previous American experience.

Saudi Arabia is three times the size of Texas, largely desert, and located astride the southern portion of what today are estimated to be more than half the petroleum reserves of the noncommunist world.[3] The largest portion of those reserves are in

---

[1] See, for example, memorandum, "The Problem," March 13, 1944, folder: "Oil," Duffield Office Files, Record Group 80, General Records of the Department of the Navy (hereafter cited as RG80). The memorandum states that the decline began in 1939, but the American Petroleum Institute's historical records show annual discoveries below annual production only in 1942 and 1943: *Petroleum Facts and Figures, 1971 Edition*, p. 115.

[2] The currently preferred form of the king's name, 'Abd al-'Aziz, will be used throughout this narrative rather than Ibn Saud, the historically more common form in American documents.

[3] U.S. Congress, Senate, Committee on Interior and Insular Affairs, *Geopolitics of Energy*, by Melvin A. Conant and Fern R. Gold, p. 33.

Saudi Arabia itself, but the geological formation is centered on the
Persian Gulf[4] and is shared on the north with the sheikhdom of
Kuwait, the traditionally rival Arab state of Iraq, and the non-
Arab, but Muslim, state of Iran. To the east of Saudi Arabia, the
petroleum formation runs through the diminutive Arabic sheikhdoms of Bahrain, Qatar, Abu Dhabi, and Dubai, and the sultanate
of Oman. Besides endowing it with oil, geography conspired to
make Saudi Arabia custodian of the holy Muslim cities of Mecca
and Medina, and located it almost adjacent to the nascent state
of Israel, anathema to all Arabs. Seven hundred miles to the north,
across Iran, lies the Soviet Union, home of atheistic communism
and supporter of socialist radicals throughout the Middle East.
One could hardly have found a more volatile location for oil
reserves critical to the industrial West.

The monarch of this realm, 'Abd al-'Aziz, had roots in a tradition quite unlike any in the West. Present-day Saudi Arabia was
his personal creation, forged with great astuteness, courage, and
religious fervor during thirty years of hard-fought desert campaign
and political maneuver.[5] The loyalty of his subjects was derived
from concurrent roles as imam (religious leader) of the Wahhabis,
and malik (king in the modern sense of the term).[6] And the kingdom that he established rested on foundations laid in the eighteenth
century by his ancestor, Muhammad ibn Sa'ud, an ardent follower
of the Sunnite Muslim reformer Muhammad ibn 'Abd al-Wahhab.[7]
'Abd al-'Aziz' approach to the West is best understood in the
historical context from which it arose.

    [4] Sa'udi Arabs call the body of water the Arabian Gulf, but Persian Gulf is
more common and will be used here.
    [5] There is a lively account of the career of 'Abd al-'Aziz in Jacques Benoist-
Méchin, *Arabian Destiny*. The standard source for 'Abd al-'Aziz' early career is
H[arold] C. Armstrong, *Lord of Arabia*. A good, concise treatment of the subject
is in the *Aramco Handbook: Oil and the Middle East*, pp. 49-65. The classic
sympathetic account is H[arry] St. John [Bridger] Philby, *Sa'udi Arabia*, pp. 237-
358. For the larger Islamic context of these events, see H.A.R. Gibb, *Mohammedanism: An Historical Survey*.
    [6] George A. Lipsky and others, *Saudi Arabia: Its People, Its Society, Its Culture*,
pp. 106-129.
    [7] George Rentz, "Wahhabism and Saudi Arabia," in Derek Hopwood, ed.,
The Arabian Peninsula: Society and Politics, pp. 54-66.

Since shortly after its founding in the seventh century, Islam has been divided into an orthodox Sunnite majority and a doctrinally more heterogeneous Shi'ite minority concentrated mainly in Iran, Iraq, and Yemen. But by the seventeenth century Sunnite orthodoxy had lost much of its vitality to legalism, mysticism, and practices bordering on idolatry. It was reaction to these trends and a desire to return to an intense, uncomplicated faith that inspired the revival and reform movement of Muhammad ibn 'Abd al-Wahhab in central Arabia in the early part of the eighteenth century.[8] His followers have since been known as Wahhabites in the West, although they themselves shun the term in favor of Muwahhidun (Unitarians). The movement was a call for return to the doctrine of the oneness of God and the strict and simple teachings of the Quran and the Prophet Muhammad.

One of the earliest converts to this movement was Muhammad ibn Sa'ud, ruler of the town of al-Dir'iyah, just north of al-Riyad, the present capital, in the Najd region of central Arabia. In 1744 Ibn Sa'ud and Ibn 'Abd al-Wahhab bound their families into a compact of mutual support that has lasted for more than two centuries. The combination proved a formidable one. Arabic society, and especially Bedouin society, has placed high value on kinship, personal relationships, courage, and fierce independence.[9] In the eighteenth century this translated into a society of constant warfare between autonomous desert tribes and villages, with a semblance of order maintained only on the western and eastern fringes by the nominal rule of the Ottoman Turks. Wahhabism provided the cohesion of loyalty to a larger cause, and by military conquest and religious conversion during the next fifty years the House of Sa'ud gradually spread its control over most of the interior of the Arabian Peninsula. By the early nineteenth century, the movement became a matter of concern to the Ottoman Turks. In 1802 Wahhabite forces sacked the Shi'ite holy city of Karbala in Iraq, and in 1803 they took control of Mecca in western Arabia and proceeded to destroy all the stringed instruments they

[8] Except as otherwise noted, this account of the early Wahhabite movement is based on Rentz, "Wahhabism," in Hopwood, pp. 54-66.

[9] Lipsky, pp. 2, 47, 64-79.

could find, and all the shrines dedicated to the worship of saints rather than God. Two years later they performed a similar cleansing operation in Medina.

The pasha (viceroy) of Egypt, Muhammad 'Ali, was directed to restore Ottoman control, and in 1811 he launched the first in a long series of campaigns to accomplish that. Finally, in 1818, he forced the surrender of 'Abd Allah, the head of the House of Sa'ud, and the following year he razed the city of al-Dir'iyah to the ground. After three-quarters of a century, the Wahhabite state was reduced to a tenuous hold on portions of central Arabia. Wahhabism retained its vitality, but the House of Sa'ud suffered from severe internal dissension; by the end of the nineteenth century it had been supplanted by the House of Rashid as the dominant power in the Najd. It was this situation that 'Abd al-'Aziz sought to rectify in 1902.

As a youth, 'Abd al-'Aziz had absorbed a strict Islamic viewpoint from his father, a thorough knowledge of Bedouin ways from flight through southern Najd, Qatar, and Bahrain, and respect for survival by manipulation of great power rivalry from almost a decade of exile in the tiny sheikhdom of Kuwait.[10] He was a man of immense physical strength, vitality, and courage, and at twenty-one he decided to focus this energy on restoration of the House of Sa'ud. His first stroke was a bold one. After several months of unsuccessful attempts to rally southern tribes to his cause, he decided to strike directly at the stronghold of his enemy. With a tiny band of fifty men, he slipped by night into the heavily defended city of al-Riyad, seized the inner fortress, killed the local governor, and announced the return of the House of Sa'ud. He brought his father, who still carried the title "Imam of the Wahhabites," home from exile in Kuwait, and with his father's blessing he took control of military and political affairs. Al-Riyad was only one city, however, and it took four years of heavy desert fighting to take control of the rest of central Arabia from the House of Rashid.

---

[10] Except as otherwise noted, this account of the rise of 'Abd al-'Aziz is based on Benoist-Méchin, pp. 57-181.

'Abd al-'Aziz now faced the traditional Arab problem—how to weld the volatile Bedouin tribesmen into a reasonably cohesive unit. After several years of recurring internal and external conflict, he turned in 1912 to a system of religious-military-agricultural colonies called the Ikhwan (Brotherhood).[11] Bedouins were recruited for the colonies by zealous missionaries, and their attention was focused on religious instruction and agricultural pursuits. In time, over a hundred of these Ikhwan colonies were established, and their members became intense in both their religious fervor and their military zeal. In effect, 'Abd al-'Aziz breathed new life into Wahhabism as a cohesive force among the townsmen and Bedouin tribesmen. Ikhwan fighting men contributed immensely to the continuing campaign to reunify the Arabian Peninsula.

World War I involved Arabia in British attempts to wrest the area from the Ottoman Turks, but, unfortunately, the British were of two minds. The Persian Gulf lay within the jurisdiction of the British India Office and it was through this channel that support reached 'Abd al-'Aziz. It was a mission for the India Office that in 1916 first brought to Arabia H. St. John B. Philby, later convert to Islam, confidant of 'Abd al-'Aziz, and student of the region. The Red Sea and Hijaz region, however, were the responsibility of the War Office and its Arab Bureau in Cairo, the best known representative of which was T. E. Lawrence, "Lawrence of Arabia." The dominant Arab figure in the Hijaz was al-Husain ibn 'Ali, Ottoman-appointed Sharif of Mecca, descendant of the Prophet Muhammad's own family, the Hashimites, and bitter rival of 'Abd al-'Aziz for Arab loyalties. British subsidies and promises of postwar support led al-Husain to raise the flag of revolt against the Ottoman Empire in 1916—a revolt in which 'Abd al-'Aziz refused to participate actively despite his strong opposition to the Turks. A further byproduct of this period was a series of British agreements with Arab rulers (including 'Abd al-'Aziz) not to reach political or commercial agreements with non-British subjects without British consent. The British, in turn, provided a degree of

[11] Lipsky, p. 12, states that the Ikhwan movement began in the nineteenth century and was taken over by 'Abd al-'Aziz in 1916, but this account follows Benoist-Méchin, pp. 116-121, and Philby, *Sa'udi Arabia*, p. 261.

protection and an annual subsidy. The agreement with 'Abd al-'Aziz was ended in 1927, but those with the rulers of Bahrain and Kuwait continued into the 1930s.

When the war was over, al-Husain was recognized by the British as king of the Hijaz, though he himself assumed the title "King of the Arabs." Al-Husain's son Faisal was first proclaimed king of an independent Syria, but his refusal to accept the French mandate led him to be forced into exile, and he subsequently became king of British-mandated Iraq. 'Abd al-'Aziz thus found himself surrounded on the west and on the north by the rival Hashimite family, and his previous confidence in British support cooled considerably. In the years immediately after the war, 'Abd al-'Aziz and his Ikhwan fighting men steadily expanded Sa'udi control in the north to the present border with Iraq and Jordan, and in the southwest to shores of the Red Sea south of Mecca. Mecca, Medina, and the Hijaz, however, remained in control of al-Husain and the British-supported Hashimite family.

In 1924 the Republic of Turkey abolished the caliphate and al-Husain immediately had himself declared caliph of all Islam. This, to the Wahhabites, was the final indignity in a long line of indignities, and 'Abd al-'Aziz finally unleashed the Ikhwan against the Hashimites. In a swift campaign the Wahhabite forces drove straight to Mecca; al-Husain abdicated in favor of his son 'Ali and fled into exile; and 'Ali surrendered after several months of seige in Jiddah. In 1926 'Abd al-'Aziz was proclaimed king of the Hijaz and afterwards also king of Najd and its dependencies. Consolidation of this realm was no easy task, however. In 1928 several Ikhwan leaders defied orders and carried out raids in British-mandated Iraq and Transjordan. Attempts to discipline them led to fierce internal fighting, and the rebel leaders were not defeated until 1930. On September 22, 1932, the country was renamed the "Kingdom of Saudi Arabia."

In the thirty years since his raid at al-Riyad, 'Abd al-'Aziz had forged a desert kingdom of townsmen and mercurial Bedouin tribesmen, established himself as keeper of the holy cities of Mecca and Medina, and developed a deep distrust of the British and their Hashimite allies in Iraq and Jordan. He was a man of broad knowledge, great insight, and iron will. His problem now was to consolidate control. He had built a reputation for ruthlessness to those who continued to oppose him, benevolence to his friends and those who acknowledged his authority, and astuteness in understanding the politics of the region. To the loyalties he already derived from roles as religious leader and temporal king, 'Abd al-'Aziz added the bond of kinship ties. Taking advantage of the Muslim custom of four wives and relatively easy divorce, he had, over the years, married the daughters of numerous tribal leaders, fathered thirty-one sons,[12] and created strong blood ties with the most powerful tribes in Saudi Arabia. In addition, he used liberal financial subsidies to prevent the Bedouins from raiding one another and to keep them loyal to himself. This practice caused concern, because, as Philby remarked, 'Abd al-'Aziz' income never matched ''his generous conception of his functions and obligations as a ruler.''[13] He rarely saw a newspaper, but he had translators monitor foreign broadcasts and keep him continuously abreast of developments abroad.[14]

The country was an impoverished desert, dependent in large measure on income from the annual pilgrimage to Mecca and on customs dues for revenue, but the worldwide depression affected Muslims, too, and revenue from these sources had seriously declined in the early 1930s. It was clear that sources of additional revenue were imperative, and it was in this context that news reached al-Riyad of the discovery of oil in the adjacent sheikhdom of Bahrain by the Bahrain Petroleum Company, a subsidiary of Standard Oil of California, in June 1932. Sporadic interest in the possibility of oil in the al-Hasa region dated back to 1922, but as late as 1930 'Abd al-'Aziz had turned down a Socal request for permission to conduct geological explorations in the area.[15] The Bahrain discovery changed his mind. Oil as a possible source of revenue and a buttress to his regime now overrode his reluctance

---

[12] Benoist-Méchin, p. 289.   [13] Philby, *Sa'udi Arabia*, p. 333.
[14] Richard H. Sanger, *The Arabian Peninsula*, p. 37.
[15] *Aramco Handbook*, p. 108.

to permit non-Muslim foreigners to develop the area, and 'Abd al-'Aziz embarked on a new venture.

*

ALTHOUGH petroleum may have been a new commodity for 'Abd al-'Aziz in 1932, it had already become deeply imbedded in Middle Eastern politics and British policy. The story is rather complex, but it is worth reviewing in some detail because it provides the background for much of what occurred between 1943 and 1947.[16] Surface indications of petroleum had been known throughout the Middle East since the beginning of recorded history, but serious interest in its development did not begin until after the American and Russian petroleum industries emerged in the latter part of the nineteenth century. Production came first in Persia (renamed Iran in 1935) as a result of the work of an adventurous Australian miner, William Knox D'Arcy. In 1901 D'Arcy was offered and accepted a sixty-year concession covering most of Persia. The initial drilling was unsuccessful, but in 1905 the British Naval Fuel Committee induced the British-owned Burmah Oil Company to form a syndicate to finance continuing work on the concession. These funds were also exhausted by 1908, just as a final drilling attempt brought in the first producing well. This success led di-

---

[16] The following account of early competition for Middle Eastern oil is based on Stephen Hemsley Longrigg, *Oil in the Middle East: Its Discovery and Development*, pp. 16-47 and 66-70; Marian Kent, *Oil and Empire: British Policy and Mesopotamian Oil, 1900-1920*; George Sweet Gibb and Evelyn H. Knowlton, *The Resurgent Years, 1911-1927*, pp. 278-318; U.S. Congress, Senate, Select Committee on Small Business, Subcommittee on Monopoly, *The International Petroleum Cartel*, pp. 45-67; U.S. Congress, Senate, Special Committee Investigating Petroleum Resources, *Diplomatic Protection of American Petroleum Interests in Mesopotamia, Netherlands East Indies, and Mexico*, by Henry S. Fraser, pp. 1-20; and "Memoirs of Calouste Sarkis Gulbenkian, particularly concerning the history of the Iraq Petroleum Company Limited, September 16, 1945," enclosed in letter, John C. Wiley (Lisbon) to Secretary of State, March 4, 1948, file 890F.6363/3-448, Record Group 59, General Records of the Department of State (hereafter cited as RG59). Other useful accounts include Benjamin Shwadran, *The Middle East, Oil, and the Great Powers*, pp. 195-252; and George W. Stocking, *Middle East Oil: A Study in Political and Economic Controversy*, pp. 40-65.

rectly to the formation in 1909 of the Anglo-Persian Oil Company (later British Petroleum) to develop the Persian fields. All of this was of great interest to First Lord of the Admiralty Winston Churchill, who wanted a reliable source of supply for the Royal Navy, and who by 1914 persuaded the British government to purchase a 51 percent interest in Anglo-Persian. The British government already had a strong interest in protecting its lines of communication to the rest of the empire from German and Russian incursions, and oil now provided an additional reason for wanting to maintain a secure position in the Middle East.[17]

Meanwhile, another company under partial British ownership had emerged and begun to show interest in the area. In 1903 Henri Deterding, managing director of Royal Dutch,[18] with extensive production in the Netherlands East Indies, had joined with Marcus Samuel, of the British Shell Trading and Transport Company, and the French Rothschild interests to form a jointly owned oriental marketing organization, the Asiatic Petroleum Company. This worked so well that the unique Royal Dutch-Shell alliance was formed in 1907. Royal Dutch and Shell were converted into holding companies, sharing ownership on a 60 percent/40 percent basis in three operating subsidiaries: Asiatic for marketing, N. V. De Bataafsche Petroleum Maatschappij for production, and the Anglo-Saxon Petroleum Company, Ltd., primarily for transportation. Deterding provided the driving force for this new organization until his retirement in the 1930s, and among his many early interests were new sources of supply in the Middle East in general and the Ottoman Empire in particular.

Events in the Ottoman Empire were far more complex than in

[17] For an overview of the British viewpoint, see Elizabeth Monroe, *Britain's Moment in the Middle East, 1914-1956*, pp. 95-115. The author is indebted to Ms. Monroe for a long conversation on this subject in London, in May 1977.

[18] The full name of the company was Naamlooze Vennootschap Koninklijke Nederlandsche Maatshappij tot Exploitatie van Petroleumbronen in Nederlandsche-Indië. The origins of Royal Dutch are described in considerable detail in F. C. Gerretson, *History of the Royal Dutch*. A concise account of the origins of both Royal Dutch and Shell is contained in Kendall Beaton, *Enterprise in Oil: A History of Shell in the United States*, pp. 20-39; and Harold F. Williamson et al., *The American Petroleum Industry*, vol. 1, pp. 664-74.

Persia. Production there did not begin until 1927, but it was preceded by thirty years of competition for the right to participate in it. Among the early competitors was a young London-educated Armenian civil engineer, Calouste Sarkis Gulbenkian, whose father was a wealthy merchant and dealer in Russian petroleum.[19] Gulbenkian had prepared a comprehensive report for the Turkish government in the 1890s on oil prospects in Mesopotamia, and he was well versed in Turkish politics. By 1910 he was a director in British-backed National Bank of Turkey, and in 1912 he was instrumental in joining that bank with Royal Dutch-Shell and the German Deutsche Bank in establishment of the Turkish Petroleum Company (TPC) for the purpose of petroleum development throughout the Ottoman Empire.[20] This action aroused concern in the British Foreign Office, which then brought pressure to bear for participation by Anglo-Persian. The upshot was a reorganization of Turkish Petroleum in 1914, with Anglo-Persian acquiring a 50 percent interest, Anglo-Saxon (Shell) 25 percent, and the Deutsche Bank 25 percent. Over his strong objections, Gulbenkian's personal share was recognized only by a lifetime 5 percent interest without voting rights, to be provided jointly by Shell and Anglo-Persian.[21] The participants pledged in a self-denying agreement not to involve themselves in oil development elsewhere in the Ottoman Empire except in association with the company. TPC held some vague German claims dating back to 1888 and 1904, and the British and German ambassadors now approached the Turkish government for a definitive concession agreement. A letter from the grand vizier to the two ambassadors dated June 28, 1914, promised leases in the vilayets of Baghdad and Mosul, but general European war broke out in August, and no formal concessionary agreements were signed.

World War I added additional levels of complexity. The Ottoman Empire came to an end, and at the Allied conference at

---

[19] "Gulbenkian Memoirs," 890G.6363/3-448, RG59.

[20] Longrigg, pp. 27-30. The author is indebted to Brigadier Longrigg for a highly informative converstion on this subject at his home in Surrey, England, in May 1977.

[21] Kent, pp. 170-71, gives the text of this agreement.

San Remo, Italy, in April 1920, its Arab territories were trans-
formed into the French mandates of Lebanon and Syria and the
British mandates of Mesopotamia (Iraq) and Palestine (including
Palestine and Transjordan).[22] As a result of wartime negotiations
instigated by Gulbenkian and Deterding,[23] an agreement regarding
disposition of the German interest in the Turkish Petroleum Company was also worked out and officially ratified at San Remo. The
Deutsche Bank holdings were to be transferred to a new French
company, eventually the Compagnie Française des Pétroles
(C.F.P.), organized specifically for this purpose. Ownership in
the Turkish Petroleum Company was now divided among Anglo-
Persian (47.5 percent), Anglo-Saxon (22.5 percent), the French
(25 percent), and Gulbenkian (5 percent).[24] The self-denying
agreement of 1914 continued in effect. The company had as yet
no firm concession, much less a producing well, but it held the
Turkish grand vizier's "promise" of a concession in the Baghdad
and Mosul areas, and the British mandate over Mesopotamia
clearly gave it an advantage in requesting that the successor government honor the pledge. This was enough to bring cries of
"foul" from the United States, even though it had not been among
those powers at war with Turkey, and it had not ratified the
Versailles Treaty and Covenant of the League of Nations under
which the mandates had been established.

American interest in Middle Eastern oil requires some explanation, since it was a combination of postwar concern over an oil
shortage, a drive by Standard Oil (New Jersey) for additional
sources of supply, and State Department defense of equality of
commercial opportunity in independent territories. The general
climate was set by the experience of mobilizing to meet American
and Allied petroleum requirements during World War I, and

---

[22] U.S. Congress, *Diplomatic Protection of American Petroleum Interests*,
p. 2.
[23] "Gulbenkian Memoirs," 890G.6363/3-448, RG59; and Longrigg, p. 44.
Kent (pp. 137-57) gives a detailed account of these negotiations from the Foreign
Office viewpoint, and does not mention Gulbenkian's role. He either operated
behind the scenes or overstated his importance in his memoirs.
[24] Longrigg, p. 44.

highly pessimistic postwar governmental forecasts about the adequacy of American reserves. Petroleum was clearly a strategic necessity, and there were mounting demands for a worldwide search for other sources of supply. In mid-1919 the Department of State began instructing its officers overseas to lend full support to this drive.[25] The "oil panic" lasted until about 1924, when new discoveries in Oklahoma, Texas, and California not only removed the threat of exhaustion but actually produced a surplus. The San Remo conference, however, fell at the peak of this national concern, and the idea of being excluded from Mesopotamia was not well received.

The concern took more concrete form in the mind of Walter C. Teagle, who had assumed the presidency of the Standard Oil Company (New Jersey) in 1917, and who by 1920 had already been thwarted by British mandate authorities in his attempts to send exploratory parties into Mesopotamia.[26] Jersey had been the central holding company in John D. Rockefeller's original Standard Oil Company, and when the Supreme Court in 1911 ordered it to divest itself of over thirty-three subsidiaries (including Standard Oil of New York and Standard Oil of California), Jersey had been left with a serious supply problem. Rockefeller's success had convinced the industry of the competitive strength to be gained from fully integrated operations, with central management control over all elements of the business—production, transportation, refining, and marketing. The dissolution left Jersey with direct command over only 8 percent of the production required for its refinery operations,[27] and rectification of this imbalance was one of the major tasks undertaken by Teagle when he assumed the presi-

[25] On the post-World War I "oil scare," see John A. Denovo, "The Movement for an Aggressive American Oil Policy Abroad, 1918-1920"; Gerald A. Nash, *United States Oil Policy, 1890-1964: Business and Government in Twentieth Century America*, pp. 23-48; and U.S. Congress, Senate, Special Committee Investigating Petroleum Resources, *American Petroleum Interests in Foreign Countries, Hearings*, pt. 3 of 6, pp. 1-27, 297-383.

[26] Gibb and Knowlton, pp. 284-88.

[27] Refinery throughput in 1912 was 96,000 barrels/day, and the combined production of Standard Oil of Louisiana, Carter Oil, and a few other subsidiaries was only 7,500 barrels/day; Gibb and Knowlton, p. 44.

dency.[28] Teagle's initial focus was on the Netherlands East Indies, but in February, 1919, he requested that a study be made of the possibilities in Mesopotamia.[29] Corporate attempts to obtain permission for a geological survey of the area were rebuffed by British occupation authorities, despite irrefutable evidence that such permission had been granted to Shell and Anglo-Persian. In October 1919, Jersey appealed to the Department of State for support, and an exchange between the department and the British Foreign Office on the subject of equality of treatment was already under way when news of the San Remo agreement broke in 1920.[30]

The response to Jersey's request for State Department support was prompt, forceful, and continuing. As an organization, the Department of State has always seen its mission as defense of American interests overseas, with most debate revolving around definition of those interests and design of a strategy that had some hope of success. At the time of San Remo there was consensus in American government and business circles that the national interests included access to overseas petroleum reserves, and State was quick to respond. In regard to strategy, the documents of the period show that the department was much enamored of negotiation, appeal to the legal sanction of treaties, and defense of the principle of the "Open Door" in commercial matters. Negotiation and legal sanctions were easy to understand from an organization with only those weapons under its direct control,[31] but the "Open Door" requires a closer look. In its simplest form the "Open Door" meant equality of commercial opportunity for nationals of all countries in all parts of the world, and it first took official form in Secretary of State John Hay's diplomatic notes on American

---

[28] Ibid., pp. 105-109, 278.   [29] Ibid., p. 287.

[30] Ibid., pp. 285-88; U.S. Congress, *International Petroleum Cartel*, pp. 51-52.

[31] It has been argued that, on questions of national strategy, components of the American government tend to support strategies that utilize their special competence and in which their personnel have a deep professional interest; Morton H. Halperin, *Bureaucratic Politics and Foreign Policy*, pp. 26-62. This tendency would appear to be one source of the excessive legalism in twentieth-century diplomacy commented upon by George F. Kennan in *American Diplomacy, 1900-1950*, pp. 79-89.

China policy in 1899 and 1900. As critics of American policy have abundantly noted, this principle did, in fact, support American economic expansion during much of the twentieth century.[32] But its roots were much more profound, and its invocation stemmed from more than purely commercial considerations. From its beginnings in the liberal democratic theory of John Locke and the laissez-faire economics of Adam Smith, American political and economic ideology has been grounded in the notion that maximum collective good will result from a society structured to permit freedom for individuals to compete in pursuit of their individual interests.[33] Although not always translated into practice, this core belief in the efficacy of equality of political and economic competitive opportunity has been repeatedly invoked by both defenders and critics of American society as self-evident truth.[34] At the insistence of Woodrow Wilson, the principle of equal commercial opportunity in mandated territories had been written into the Covenant of the League of Nations,[35] and this added a quasi-legal sanction in 1920. Furthermore, the department took the

[32] See, for example, William Appleman Williams, *The Tragedy of American Diplomacy*; Gabriel Kolko, *The Roots of American Foreign Policy: An Analysis of Power and Purpose*; and Harry Magdoff, *The Age of Imperialism: The Economics of U.S. Foreign Policy*.

[33] John Locke, "Second Treatise on Civil Government" (first published in 1690), in John Locke, *Locke on Politics, Religion, and Education*; and Adam Smith, *An Inquiry into the Nature and Causes of the Wealth of Nations* (first published in 1776). For discussion of the continuing American emphasis on the value of freedom for individual competition in both politics and economics, see William Ebenstein, *Modern Political Thought: The Great Issues*, pp. 127-272, 487-580; and Richard Hofstadter, *The American Political Tradition and the Men Who Made It*. N. Gordon Levin, Jr., has argued that Woodrow Wilson was the epitome of this integrated political and economic world view; see his *Woodrow Wilson and World Politics: America's Response to War and Revolution*, pp. 13-49.

[34] Readers immersed in the American political tradition may want to note the hierarchal society of Confucian China, the elitism of Edmund Burke, and the "democratic centralism" of Lenin, among others, to recall that the efficacy of individual freedom has not been a universally accepted truth. For a concise description of these other viewpoints, see John King Fairbank, *The United States and China*, pp. 15-105; and Ebenstein, pp. 275-484.

[35] Levin, pp. 245-47.

position that the principle of equal commercial opportunity applied at home as well, and consistently refused to support one American company over another in its dealings abroad. Invocation of this "self-evident" principle in this case protected the organization against charges of preferential treatment.[36] State's response to requests for support in the 1920s and again in the 1940s was clearly influenced by this deeply ingrained organizational world-view.

When details of the San Remo agreement were made public in July of 1920, Mesopotamian oil became an international cause célèbre. In a sharp diplomatic exchange Secretary of State Bainbridge Colby charged the British with violation of the mandate and challenged the validity of Turkish Petroleum's 1914 "promise" of a concession in Mesopotamia. British Foreign Secretary Lord Curzon countered with assertions that San Remo did not exclude other powers, that mandate rules were not for debate with powers that had not adhered to the league, and that Turkish Petroleum's claim was a valid one.[37] Forceful American diplomatic pressure continued for two years, but the impasse was actually resolved by private negotiations. Recognizing State's willingness to support American oil companies in general but not Jersey in particular, Teagle assembled a group of seven companies that in November 1921 notified the new secretary of state, Charles Evans Hughes, of its interest in sending a joint geological expedition to Iraq. The original seven were Jersey, Standard Oil of New York, The Texas Company, the Gulf Oil Corporation, the Sinclair Consolidated Oil Company, the Atlantic Refining Company, and the Pan American Petroleum and Transport Company.[38] The prospect of such formidable competition appalled Gulbenkian, who convinced British Undersecretary for Foreign Affairs Sir William

[36] Other nations, including the British, were not as reluctant to advance the cause of a particular company, and records of the British Foreign Office reveal a much closer working relationship with the oil companies than that which existed in the United States.

[37] U.S. Congress, *Diplomatic Protection of American Petroleum Interests*, pp. 7-19.

[38] Gibb and Knowlton, pp. 292-93. Texaco and Sinclair subsequently dropped out.

Tyrell that it was in Britain's national interest to persuade the British companies to let the Americans participate in Turkish Petroleum rather than operate independently in Iraq.[39] Tyrell agreed, and on his advice, the British companies opened discussions with the American group. By June of 1922 preliminary agreement had been reached and approved by Secretary of State Hughes, providing that the final agreement permitted participation by any reputable American company desiring it, and providing that it did not violate the principle of the "Open Door" in mandated territories.[40]

Negotiations dragged on for six years, during which time Turkish Petroleum obtained a firm concession for most of Iraq and brought in its first producing well.[41] Final agreement was reached on July 31, 1928, with ownership in Turkish Petroleum divided as shown in Table I-1. Turkish Petroleum was to be a nonprofitmaking supplier of crude to its constituents, who would purchase Gulbenkian's share of production at a fair market price, and Anglo-Persian was to receive a 10 percent overriding royalty in return for transferring half of its shares to the American group. To satisfy the Department of State's insistence on an "Open Door," plots within the TPC concession were to be made available annually for sale by competitive bidding under procedures that later proved to be unworkable. The most important provision, however, was one adopted at the insistence of Gulbenkian and the French. This was an agreement to retain the original self-denying clause of 1914 and to refrain from obtaining concessions or purchasing oil independently in any part of what was construed to have been the old Ottoman Empire. For the purpose of the agreement, this was defined by a French map on which a red line was drawn around Turkey, Iraq, Syria, Lebanon, Transjordan, Palestine, Cyprus,

---

[39] "Gulbenkian Memoirs," 890G.6363/3-448, RG59. For a persuasive argument that Gulbenkian did, in fact, play the decisive catalytic role in breaking the impasse, see Gibb and Knowlton, p. 291.

[40] U.S. Congress, *Diplomatic Protection of American Petroleum Interests*, pp. 20-21; and U.S. Congress, *International Petroleum Cartel*, pp. 52-53.

[41] Gibb and Knowlton, pp. 294-305.

TABLE I-1

DIVISION OF OWNERSHIP IN THE TURKISH PETROLEUM COMPANY[1] UNDER THE
RED LINE AGREEMENT OF JULY 31, 1928

Percent Ownership	Company
23.75	D'Arcy Exploration Co., Ltd. (subsidiary of the Anglo-Persian Oil Company, in which the British government held a 51% interest).
23.75	Anglo-Saxon Petroleum Co., Ltd. (subsidiary of Royal Dutch Shell, which was owned 60% by Royal Dutch and 40% by Shell Transport and Trading).
23.75	Compagnie Française des Pétroles (owned 35% by the French Government).
23.75	Near East Development Corporation (organized in 1928 for participation in TPC, with ownership as follows: 25% Standard Oil Company [New Jersey] 25% Standard Oil Company of New York 16.66% Gulf Oil Corporation 16.66% The Atlantic Refining Company 16.66% Pan American Petroleum and Transport Company [subsidiary of Standard Oil of Indiana])[2]
5	Participation and Investments Company (wholly owned by C. S. Gulbenkian)

SOURCE: Gibb and Knowlton, p. 306; U.S. Senate, *International Petroleum Cartel*, pp. 45 and 65.
1. In 1929 Turkish Petroleum changed its name to Iraq Petroleum Company.
2. In 1931 Standard of New York was renamed Socony-Vacuum, and together with Jersey it bought out the interests of Atlantic and Pan American, leaving Near East owned 41.66 by Jersey, 41.66 by Socony-Vacuum, and 16.66 by Gulf. Gulf withdrew in 1934, leaving only Jersey and Socony-Vacuum.

and all of the Arabian Peninsula except the sheikhdom of Kuwait.[42]

To keep all of this in perspective, it should be noted that ten years later, in 1938, Iraq was producing only 1.6 percent of the world's oil, Iran was producing 3.8 percent, and Saudi Arabia only a trace; 60 percent of the balance came from the United States, and the rest came primarily from Venezuela, Indonesia, the Soviet Union, and Romania (Table I-2). But this long, in-

[42] Gibb and Knowlton, pp. 306-308; Longrigg, pp. 69-70; U.S. Congress, *International Petroleum Cartel*, pp. 53-67.

TABLE 1-2
WORLD CRUDE OIL PRODUCTION, 1938-1965
(Thousands of metric tons and percent of Non-Communist total)

	1938		1950		1955		1960		1965	
United States	164,107	(69 7)	266,709	(55 9)	335,746	(48 8)	347,976	(39 3)	384,946	(31 0)
Venezuela	27,504	(11 7)	78,235	(16 4)	113,041	(16 5)	149,372	(16 8)	182,409	(14 7)
Other Western Hemisphere	16,799	(7 1)	27,801	(5 8)	47,429	(6 9)	70,731	(8 0)	97,757	(7 9)
Total Western Hemisphere (North America & Latin America)	208,410	(88 5)	372,745	(78 1)	496,216	(72 2)	568,078	(64 1)	665,112	(53 6)
Saudi Arabia	67	(—)	26,617	(5 6)	47,536	(6 9)	62,068	(7 0)	101,033	(8 1)
Iran	10,359	(4 4)	32,259	(6 8)	16,356	(2 4)	52,392	(5 9)	93,454	(7 5)
Iraq	4,363	(1 9)	6,650	(1 4)	33,681	(4 9)	47,281	(5 3)	64,473	(5 2)
Kuwait (inc. Neutral Zone)	—		17,291	(3 6)	56,053	(8 2)	89,157	(10 1)	128,394	(10 3)
Other Middle East [1]	1,133	(0 5)	3,159	(0 7)	7,120	(1 0)	10,972	(1 2)	29,239	(2 4)
Total Middle East	15,922	(6 8)	85,976	(18 1)	160,746	(23 4)	261,870	(29 5)	416,593	(33 5)
Libya	—		—		—		—		58,772	(4 7)
Algeria	—		3	(—)	58	(—)	8,632	(1 0)	26,025	(2 1)
Nigeria	—		—		—		850	(0 1)	13,538	(1 1)
Indonesia	7,398	(3 1)	6,673	(1 4)	12,264	(1 8)	20,844	(2 4)	23,925	(1 9)
Other Eastern Hemisphere	3,716	(1 6)	11,640	(2 4)	17,883	(2 6)	26,029	(2 9)	38,305	(3 1)
Total Eastern Hemisphere (Non-Communist)	27,036	(11 5)	104,292	(21 9)	190,951	(27 8)	318,225	(35 9)	577,158	(46 4)
Total Non-Communist	235,446	(100 0)	477,037	(100 0)	687,167	(100 0)	886,303	(100 0)	1,242,270	(100 0)
Communist Countries [2]	37,970		45,045		86,860		170,564		268,838	
Total World	273,416		522,082		774,027		1,056,867		1,511,108	

SOURCE: Joel Darmstadter, Perry D Tetelbum, and Jaroslav Polach, *Energy in the World Economy A Statistical Review of Trends in Output, Trade, and Consumption Since 1925* (Baltimore The Johns Hopkins Press, 1971), pp 185-223

1. Bahrein, Qatar, Trucial Oman, Israel, Lebanon, Syria, Turkey
2 USSR, Albania, Bulgaria, Czechoslovakia, East Germany, Hungary Poland, Romania, China

volved struggle over Iraqi oil set the stage for the initial development of Saudi Arabian oil by two non-Red Line companies, Standard of California, and The Texas Company. And Gulbenkian and the Red Line Agreement proved to be the chief obstacles to be overcome in 1946 and 1947 when Jersey and Socony-Vacuum wanted to buy into the Saudi concession.

*

IT WOULD appear redundant to note that Standard Oil of California, the company that obtained the first Saudi concession, was a California company, but that point is worth emphasis. Originating in 1879 as the production-oriented Pacific Coast Oil Company, the company was purchased in 1900 by Standard Oil, converted into Standard of California in 1906 by merger into it of Standard's early West Coast marketing operations, and cut loose by the Supreme Court in 1911 as the only fully integrated successor of the original Standard Oil combine.[43] Based in San Francisco and concentrating its attention west of the Rocky Mountains, the company had a reputation for conservative self-sufficiency. Supply requirements were met chiefly by the company's own domestic production, supplemented by long-term supply contracts.[44] In 1911 it had no foreign operations—production or marketing—but like other American companies it was shaken by the post-World War I "oil scare" and it began extensive overseas explorations in the 1920s.[45] By 1928 Socal had spent approximately $50 million on ventures in Mexico, Central America, Venezuela, Colombia, the Philippines, and the Netherlands East Indies—with no success of

[43] Ralph W. Hidy, and Muriel E. Hidy, *Pioneering in Big Business, 1882-1911*, pp. 343-45; Gibb and Knowlton, pp. 8-9. For the history of Standard Oil of California through 1919, see Gerald T. White, *Formative Years in the Far West: A History of the Standard Oil Company of California and Predecessors Through 1919.*

[44] Letter to the author from W. K. Morris, vice-president, Standard Oil Company of California, March 15, 1978. The author is also indebted to George T. Ballou of Socal for a long conversation on Socal's management style in San Francisco, August 24, 1977.

[45] Interview with R. Gwin Follis, former president and chairman of the board of Standard Oil of California, in San Francisco, August 24, 1977.

any consequence.[46] The American "oil scare" was now well over, but a strong interest in foreign exploration as a hedge against the future was retained by the company geologists in general and Vice-President M. E. (Mike) Lombardi in particular. In 1928, this interest became focused on the Persian Gulf.[47]

On behalf of a London speculative company, the Eastern and General Syndicate, an energetic New Zealander, Major Frank Holmes, had obtained concessionary rights in al-Hasa in 1923, the Kuwait Neutral Zone in 1924, and the island of Bahrain in 1925.[48] The al-Hasa and Neutral Zone concessions eventually were allowed to lapse, but Eastern and General succeeded in 1927 in interesting Gulf Oil in taking an option on the Bahrain concession.[49] Before that option could be exercised, Gulf signed the Red Line Agreement, and immediately discovered that its partners, especially Anglo-Persian, would agree to neither independent nor joint operations on Bahrain.[50] Gulf had to decide whether to withdraw from Turkish Petroleum or give up the option. On the advice of Secretary of the Treasury Andrew Mellon, whose family held a dominant interest in Gulf, the company's board decided to give up the option.[51] At a meeting of the American Petroleum Institute

[46] Frederick L. Moore, Jr., "Origin of American Oil Concessions in Bahrein, Kuwait, and Saudi Arabia," p. 62. Moore's thesis was based in part on the recollections of Thomas E. Ward, who handled the American negotiations for the Eastern and General Syndicate in the 1920s. Ward's own account, including copies of numerous relevant documents, is contained in Thomas E. Ward, *Negotiations for Oil Concessions in Bahrein, El Hasa (Saudi Arabia), the Neutral Zone, Qatar, and Kuwait.*

[47] This account of the negotiations for Socal's original concession is based primarily on the memoirs of three direct participants: H[arry] St. John B[ridger] Philby, *Arabian Oil Ventures*, pp. 53-134; Karl S. Twitchell, *Saudi Arabia: With an Account of the Development of Its Natural Resources*, pp. 139-156; and the previously cited work of Stephen H. Longrigg, pp. 98-110. Concise, well-researched accounts also appear in Shwadran, pp. 301-307, and Stocking, pp. 66-84.

[48] Philby, *Arabian Oil*, p. 54; Longrigg, p. 100.

[49] Philby, *Arabian Oil*, p. 67.

[50] U.S. Congress, *International Petroleum Cartel*, pp. 71-72; Moore, pp. 57-59.

[51] Moore, p. 58.

in Chicago in 1928, Gulf officials broached the subject to Kenneth R. Kingsbury, president of Standard Oil of California, and shortly thereafter the Bahrain option was taken up by Socal.[52] Gulf went on to investigate the non-Red Line sheikhdom of Kuwait, and in 1934 it acquired a joint concession there with Anglo-Persian. Gulf also withdrew from Turkish Petroleum (now Iraq Petroleum) in 1934, leaving Jersey and Socony-Vacuum as the only American participants. The British government objected to a non-British company taking up a concession on Bahrain, but after a short exchange between the Department of State and the Foreign Office, the issue was resolved by Socal establishment of a Canadian subsidiary, the Bahrain Petroleum Company, in 1929 to hold the concession formally.[53] By the spring of 1930, Fred A. Davies, later board chairman of Aramco, arrived in Bahrain to select a site for the first test well.[54] Socal now had a foothold in the Persian Gulf—only twenty miles offshore from the Saudi mainland.

From the very first, Davies and the Socal geologists on Bahrain recommended that the company explore the mainland,[55] but matters dragged on until 1932, when the first successful Bahrain well gave added impetus to the matter. Largely at the urging of Mike Lombardi, who was charged with oversight of the Bahrain operation, Socal now began actively searching for an entrée into Saudi Arabia.[56] Its efforts intersected a chain of events coming from another direction. 'Abd al-'Aziz' intense interest in financial resources to offset the sharp decline in pilgrimage and customs income had led his confidant, H. St. John Philby, to arrange a visit to Saudi Arabia in 1931 by Charles R. Crane, a wealthy American philanthropist who had already financed a number of projects in Yemen.[57] Crane, in turn, sponsored a search for evidence of water and mineral resources in Saudi Arabia by an American mining engineer, Karl R. Twitchell, in 1931 and 1932.

[52] Ibid., p. 60; Charles W. Hamilton, *Americans and Oil in the Middle East*, p. 126.
[53] Longrigg, p. 102; U.S. Congress, *American Petroleum Interests in Foreign Countries*, p. 318; Hamilton, pp. 128-29.
[54] *Aramco Handbook*, p. 108.   [55] Ibid.; Moore, p. 78.
[56] Twitchell, p. 150.   [57] Philby, *Arabian Oil*, p. 75.

Twitchell concluded that the geology of al-Hasa was sufficiently similar to that on Bahrain to warrant serious exploration if oil were discovered on that island. The first Bahrain well came in in June of 1932, and in July the king asked Twitchell, who by that time was in New York, to locate an American company to explore his own country.[58] Among others, Twitchell approached James Terry Duce of The Texas Company (later a vice-president of Aramco), C. Stuart Morgan of Near East Development (Jersey and Socony-Vacuum), and Guy Stevens of Gulf.[59] None of these companies were interested, but at this point Twitchell was contacted by Socal, and a meeting was arranged with Lombardi in New York. It developed that Philby himself had already been contacted during a visit to London by Francis B. Loomis, another Socal vice-president,[60] and a coordinated plan of attack was now worked out to negotiate the concession. Lloyd N. Hamilton of Socal's legal staff would go to Jiddah to negotiate with 'Abd Allah Al Sulaiman Al Hamdan, the king's finance minister, and would be accompanied by Twitchell as technical advisor. Philby, on retainer to Socal, would remain in the background as advisor to both parties.[61] This was the plan put into effect in February 1933.

As negotiations proceeded, it became clear that 'Abd al-'Aziz' principal concern was the size of the cash advance that could be obtained against future royalties. Hamilton's original offer was $50,000 against the king's desire for £100,000 (at $3.30 to the £).[62] While discussions were under way to bridge this gap, Stephen H. Longrigg arrived in Jiddah as a representative of Iraq Petroleum, but IPC appears to have been more interested in a preemptive concession than in actually developing the al-Hasa area, and at any rate it was only willing to offer a cash advance of £10,000. When these positions became clear to all parties, the IPC board decided not to pursue the matter further and Longrigg retired from the scene.[63] In his detailed and charmingly candid account, Philby

[58] Twitchell, p. 148.   [59] Ibid., p. 149.
[60] Philby, *Arabian Oil*, pp. 77-78.   [61] Ibid., pp. 85-89.
[62] Ibid., pp. 89, 106.
[63] Ibid., pp. 104-120; letter, J. Skliros (IPC general manager, London) to (IPC)

reveals that 'Abd al-'Aziz' insistence on a sizable cash advance stemmed from his belief that there was little or no oil in Saudi Arabia; he was quite willing to have the highest bidder pay for the privilege of confirming that fact.[64] Despite contemporary newspaper accounts of 'Abd al-'Aziz' preferences for America's policies over those of Britain,[65] and despite his subsequent strong affinity for the United States, 'Abd al-'Aziz' decision to award the concession to Socal in 1933 appears to have been based primarily on his desperate need for revenue.

Agreement was finally reached and a concession was granted on May 29, 1933, for essentially the eastern half of Saudi Arabia. The agreement provided for an initial cash advance of £50,000, an annual "rental" of £5,000 until oil was discovered, a further cash advance of £100,000 after discovery, and royalties at the rate of 4 shillings per ton. The government agreed to forego its right to tax the company, and the company agreed "as far as practical" to employ Saudi nationals, and to refrain from interference with the internal affairs of the country.[66] Both parties obtained essentially what they wanted, and neither had any idea

---

Outside Concessions Committee, March 20, 1933, folder 32923 and minutes of IPC group representatives meetings, London, March 22 and May 5, 1933, folder 33358, corporate records acquired for *International Petroleum Cartel, Report*, Freedom of Information Branch, Federal Trade Commission (hereafter cited as FTC records); Longrigg, p. 107; U.S. Congress, *International Petroleum Cartel*, p. 74; Shwadran, p. 304; Stocking, p. 78.

[64] Philby, *Arabian Oil*, p. 133.

[65] *New York Times*, July 15, 1933, as reported in Shwadran (1955 edition), p. 291 n. 11. The idea that 'Abd al-'Aziz' original decision was based on a preference for America has persisted (see, for example, Sanger, p. 101) despite the weight of evidence that it does not appear to be true. Longrigg reports that 'Abd al-'Aziz assured *both sides* that their "company and nationality would, all things being equal, be the more acceptable" (Longrigg, p. 107), and the king apparently continued to assure the Americans of that in subsequent years.

[66] Text of the concessionary agreement of May 29, 1933, printed as annex 1 to Saudi Arabia, *Arbitration Between the Government of Saudi Arabia and Arabian American Oil Company*, Volume V, [First] *Memorial of Arabian American Oil Company*. The size of the concession was subsequently increased in 1939, and then decreased in 1947 and 1963, with further relinquishments of undeveloped areas at five-year intervals thereafter (*Aramco Handbook*, p. 112).

of the bonanza that was to follow. Socal established a wholly owned subsidiary, the California Arabian Standard Oil Company (Casoc), incorporated in Delaware, to execute this agreement. (Incorporation in the United States meant that the company's accounts were kept in dollars, rather than in sterling as in the case of Anglo-Iranian and IPC. As will be seen, this detail became a matter of great importance in the late 1940s.) It might also be noted that the American government played no role in the original Saudi concession; there was not even an American consul assigned to the country until 1942.[67]

Socal geologists arrived to begin exploratory work in September 1933, but the discovery of oil on Bahrain had meanwhile set in motion another chain of events. The Bahrain field proved to be a productive one (although much smaller than the ones found later in Saudi Arabia), and Socal now had the problem of finding a market. The company had no foreign marketing organization of its own, and attempts to find buyers for Bahrain crude foundered on several complications. Jersey and Shell were basically crude-short and interested, but they were effectively blocked by Gulbenkian and the French from buying "Red Line" oil.[68] And efforts

[67] The United States extended diplomatic recognition to Saudi Arabia (as "The Kingdom of the Hedjaz and Nejd and Its Dependencies") on May 1, 1931, signed a routine executive agreement covering "Diplomatic and Consular Representation, Juridical Protection, Commerce and Navigation" with Saudi representatives in London on November 7, 1977, accredited the American minister to Egypt concurrently minister to Saudi Arabia on August 7, 1939, and established a one-man legation and consulate in Jidda on April 22, 1942. U.S. Department of State, *Foreign Relations of the United States* (hereafter cited as *FR[date]*); *FR(1931)*, II, 551; *FR(1933)*, II, 999-1011; *FR(1939)*, IV, 830; and *FR(1942)*, IV, 560.

[68] Follis interview; letter, Guy Wellman to Teagle, May 22, 1935, folder 32923; letter, E. G. Seidel to Wellman, July 19, 1935, folder 32947; and letter, Seidel to Teagle, February 10, 1936, folder 32947, FTC records; U.S. Congress, *International Petroleum Cartel*, pp. 74-84. By 1939 IPC had worked out an internal compromise that would permit IPC to form a partnership with Socal and Texas for the development of all or part of Saudi Arabia and Bahrain, and John Brown of Socony-Vacuum was authorized to broach the idea to "Cap" Reiber of The Texas Company, but these negotiations were overtaken by World War II and dropped; minutes of (IPC) group meeting, April 5, 1939, folder 32934, and letter, J. Skliros to (IPC) groups, April 12, 1939, folder 32926, FTC records.

to sell crude direct to European refineries disclosed the fact that most of them were not equipped to handle the high sulphur content of Arabian crude.[69] By 1935 it had become clear that Socal would have to build its own refinery and create or find a marketing organization to sell the products. Both would cost money, scarce in the midst of the Depression, but Socal was determined to hold the concession and make it a profitable one. The problem of equipment for a refinery was solved by barter—Bahrain crude and California products exchanged for mostly German pipes, valves, and engines.[70] A 10,000 barrel/day refinery was put in operation by 1937, and its capacity increased to 33,000 barrel/day by the time war broke out.[71] But marketing required a longer-term solution.

Early in 1936, Kingsbury, on his way home from a discouraging London conference with Shell, happened to meet Torkild ("Cap") Reiber (board chairman of Texaco) and W.S.S. ("Star") Rodgers (president of Texaco) in a New York club, and their conversation resulted in Reiber going to San Francisco to work out a combining of interests.[72] The Texas Company had built a worldwide marketing organization to dispose of Texas production, but the lower cost of crude from high flowing Persian Gulf wells and the proximity of Bahrain to Eastern Hemisphere markets made this new source of supply quite attractive to Reiber.[73] To handle the merger, The Texas Company took a 50 percent interest in the Bahrain Petroleum Company, and all of The Texas Company's marketing facilities east of Suez were consolidated into a new subsidiary of

[69] Follis interview; Follis was a member of the Socal team that toured Europe in an attempt to sell Bahrain crude, and he was directly involved in corporate discussions of Persian Gulf matters from that point on. His recollections have been most helpful for an understanding of Socal's viewpoint.

[70] Follis interview.   [71] Longrigg, p. 103.

[72] Interview with Augustus C. Long, former board chairman of The Texas Company, in New York, August 3, 1977. Long was closely associated with corporate decision making in the 1940s, and his recollections were most helpful for an understanding of Texaco's viewpoint. For a general background on The Texas Company, see Marquis James, *The Texaco Story: The First Fifty Years, 1902-1952.*

[73] Follis interview.

Bapco—the California Texas Oil Company, Ltd. (Caltex). In a separate transaction, The Texas Company received a 50 percent interest in the California Arabian Standard Oil Company, holder of the Saudi concession. For this, The Texas Company made a payment of $3 million in cash and $18 million in deferred payments to be paid out of Arabian earnings.[74] From the Socal viewpoint, the problem of marketing Persian Gulf oil was solved by the creation of Caltex, but this was before anyone realized the quantity of oil that would be found in Saudi Arabia.[75]

Oil in commercial quantities in Saudi Arabia was discovered in March 1938, in Dammam field (near the coast 20 miles west of Bahrain and 275 miles northeast of al-Riyad).[76] Living and operating facilites began to be built at Dhahran, crude was initially barged to the refinery at Bahrain, a deep-water port was constructed at Ras Tanura, and exploration work continued to locate promising deposits.[77] Production rose from 495,135 barrels in 1938, to 3,933,907 in 1939, and 5,074,838 in 1940, but dropped back to 4,310,110 in 1941 when the war curtailed the availability of tankers to handle it.[78] By 1941 the three structures that had been tapped were estimated to have reserves of 750 million barrels, and the existence of numerous similar untapped structures

---

[74] U.S. Congress, *International Petroleum Cartel*, pp. 115-116. The marketing subsidiaries merged into Caltex were The Texas Company (Australasia), Ltd., The Texas Company (China), Ltd., The Texas Company (India), Ltd., The Texas Company (South Africa), Ltd., The Texas Company (Philippine Islands), Inc., and marketing properties in Egypt. The Texas Company also received a 50 percent interest in Socal subsidiaries holding concessions in the Netherlands East Indies and New Guinea. Socal acquired an option to buy a 50 percent interest in The Texas Company's European marketing facilities. The option was allowed to lapse in 1939, but in 1946 Caltex paid The Texas Company $28 million for those facilities. On this transaction, see also Longrigg, p. 104, Shwadran, p. 295, and Stocking, pp. 88-89.

[75] Follis interview.   [76] *Aramco Handbook*, p. 117.
[77] Ibid.

[78] Aramco crude oil production figures provided to the author by Augustus C. Long, August 3, 1977. Production continued during the war at 4,530,492 barrels in 1942, 4,868,184 barrels in 1943, 7,794,490 barrels in 1944, and 21,310,996 in 1945.

clearly indicated that potential reserves were much greater.[79] 'Abd
al-'Aziz began to recognize the value of his undeveloped resource,
and at long last he began to see an end to unrelenting financial
woes. Socal and Texaco, on the other hand, were deeply worried.
An expensive refinery would have to be built somewhere to handle
Arabian crude, the Caltex market "east of Suez" was inadequate,
and they had so far been unable to come to terms with the "Red
Line" companies that had access to the far more extensive European market. Failure to develop the reserves rapidly enough to
solve the king's financial problems could result in loss of the
concession, they thought, and at one point Socal and Texaco
considered turning back part of the concession to avoid an even
greater debacle.[80]

An early indication of the pressures to which they would be
subjected became evident in the Saudi financial crisis of 1941.
The outbreak of war had radically curtailed revenue from the
annual pilgrimage and from customs duties, and crop failures and
increasing costs of imported supplies had brought the government
to the edge of bankruptcy. Most critical was lack of funds for the

[79] U.S. Congress, Senate, Special Committee Investigating the National Defense
Program, *Investigation of the National Defense Program, Hearings*, pt. 41, *Petroleum Arrangements with Saudi Arabia*, pp. 25383-84. The following discussion
of the situation in 1941 and the attempt to secure financial aid for Saudi Arabia
is based on the extensive testimony and corporate documents in those hearings
and the government documents in *FR(1941)*, III, 624-51. These two publications
include copies of all of the relevant documents to be found in the Franklin D.
Roosevelt MSS, Franklin D. Roosevelt Library (hereafter cited as Roosevelt MSS),
and the decimal files of the Department of State at the National Archives. Additional commentary on this episode may be found in U.S. Congress, Senate,
Special Committee Investigating the National Defense Program, *Navy Purchases
of Middle East Oil*, and U.S. District Court of Southern New York, *File Civ. 39-
779* (the suit that Moffett subsequently brought against the company for services
allegedly rendered in 1941). There is an excellent analysis of this material in
Stocking, pp. 89-96; other useful summaries are in Shwadran, pp. 315-18; and
Joseph W. Walt, "Saudi Arabia and the Americans, 1928-1951," pp. 151-79.

[80] Follis interview. For comment on 'Abd al-'Aziz' shrewdness in manipulating
the companies and the great powers to achieve his own objectives, see Malcolm
C. Peck, "Saudi Arabia in United States Foreign Policy to 1958: A Study in the
Sources and Determinants of American Policy."

tribal subsidies, distributed in the form of food, clothing, and money, to retain the allegiance of the Bedouins.[81] In 1940 'Abd al-'Aziz had obtained a British subsidy of $403,000, and in 1941 this was increased to $5,285,500.[82] But by his calculations early in 1941, the king desperately needed another $6,000,000, and the companies foresaw that he would need similar amounts in the years immediately ahead, before oil revenue filled the gap. The logical solution was to request an advance against future royalties from his concessionaire, and this was the step he took in January 1941.[83] But Socal and Texaco by the end of 1940 had already advanced the king a cumulative total of $5,515,652 above royalties paid, and they had spent $30,115,241 developing the concession, without significant return at that point.[84] Money to ensure internal security was obviously critical, but the companies were reluctant to continue the "growing burden . . . of financing an independent country" in the midst of a world at war.[85] Toward the British subsidy, they had mixed feelings. Aid from that quarter would help keep 'Abd al-'Aziz in power without risking additional corporate funds. But they feared that if 'Abd al-'Aziz turned solely to the British for financial support, this action would mean increasing British influence in his government, and—in view of the past British record—a weakening of American control of the Saudi concession.[86] The logical answer then, was an approach to the American government for financial aid to Saudi Arabia.

[81] U.S. Congress, *Petroleum Arrangements with Saudi Arabia*, p. 25380. For a fuller account of the American response to the 1941 Saudi financial crisis, see Aaron D. Miller, *Search for Security: Saudi Arabian Oil and American Foreign Policy, 1939-1949*, pp. 32-48. I and others are indebted to Dr. Miller for having located and made available to researchers a large body of State Department documents pertaining to Saudi Arabia.

[82] U.S. Congress, *Petroleum Arrangements with Saudi Arabia*, p. 25381. This was the amount actually paid in 1941; when 'Abd al-'Aziz first approached the companies, he anticipated somewhat less from the British.

[83] Ibid., pp. 24804 and 25410.

[84] Ibid., pp. 25381-82.

[85] Memorandum by James A. Moffett for President Roosevelt, April 16, 1941, *FR(1941)*, III, 626.

[86] U.S. Congress, *Petroleum Arrangements with Saudi Arabia*, pp. 24829 and 25381-82.

In the January 1941 discussions Casoc president Fred Davies had promised the Saudi finance minister 'Abd Allah Al Sulaiman an advance of $3,000,000, and had agreed to attempt to persuade the parent companies to increase the loan by an additional $3,-000,000.[87] But after extensive internal consultation, the companies decided to approach the American government for the full amount.[88] Their chosen representative was James A. Moffett, oil man, former federal housing administrator, friend of Franklin Roosevelt, and in 1941 chairman of the board of both Bahrain Petroleum and Caltex. On April 9, 1941, Moffett called on Roosevelt and shortly thereafter he sent the President a formal proposal that the United States avance 'Abd al-'Aziz $6,000,000 annually for the next five years in return for that value of products provided to the American government by Casoc at a quite favorable price. The proposal also urged that the State Department seek a British commitment to increase and continue its subsidy to Saudi Arabia on a purely "political and military basis" without repayment in oil from the concession. The justification for American action was not oil, but the value of 'Abd al-'Aziz as a pro-Ally Arab leader in a highly volatile wartime area.[89] Under the original proposal, the products would have been primarily for the use of the United States Navy, and this phase of it foundered on the objection of Secretary of the Navy Frank Knox that Arabian oil had too high a sulphur content.[90] For the next few months, however, the case

[87] Ibid., pp. 25389 and 25409-11.

[88] Ibid., p. 24808.

[89] Ibid., pp. 24714-15; letter, Moffett to Roosevelt, FR(1941), III, 624-27. A further interesting clause of the proposal would have precluded resale of the products by the American government in an area that coincided with the Caltex marketing area. The companies argued that to do otherwise would have defeated the objective of providing Saudi Arabia with additional income, which was true, but the clause also carefully protected the purely commercial interests of the companies.

[90] Memorandum, Knox to Roosevelt, May 20, 1941, FR(1941), III, 635-36. Subsequent developments proved this objection to have been based on an erroneous evaluation of the sulphur problem provided to Knox by Rear Admiral H. A. Stuart, the officer in charge of naval petroleum reserves; U.S. Congress, Navy Purchases of Middle East Oil, p. 6.

for financial aid to Saudi Arabia continued to be pressed in Washington by Moffett, Max Thornburg, vice-president of Bahrain Petroleum,[91] and W.S.S. Rodgers, president and now successor to "Cap" Reiber as chief executive officer of The Texas Company. The record suggests that Rodgers was the prime mover behind this effort to solicit government funding for 'Abd al-'Aziz, and that he was motivated at least in part by the fact that Texaco at that point had invested considerably more in the Saudi concession than had Socal.

A number of proposals, including Lend Lease, an Export Import Bank loan, and a Reconstruction Finance Corporation loan were considered by the Department of State, Secretary Knox, Harry Hopkins, and Jesse Jones, Roosevelt's federal loan administrator, but in the end it was concluded that no legislative authority existed for such a loan, and that the matter could best be handled by the British.[92] Roosevelt finally ended the discussion with a note on July 18 to Jesse Jones commenting that aid to 'Abd al-'Aziz was "a little far afield" for the United States,[93] but a contemporary observer close to Jones was of the opinion that the real problem in the summer of 1941 was political concern that such a step would trigger excessive protest from midwestern isolationists.[94] Whatever the reason, overtures were made to the British by Secretary of State Cordell Hull in May and by Jesse Jones in mid-

[91] On July 7, 1941, Thornburg left Bahrain Petroleum to become special assistant to the undersecretary of state for petroleum matters. He continued as petroleum advisor until late in 1943, when he was forced to resign under charges of conflict of interest that were never formally proven.

[92] Memorandum, Jones to Secretary of State, August 6, 1941, and telegram, Secretary of State to Kirk (Egypt), August 22, 1941, *FR(1941)*, III, 642-43, and 645-46.

[93] Memorandum, Roosevelt to Jones, July 18, 1941, *FR(1941)*, III, 643. It is not clear from the record whether or not Roosevelt ever knew that—in addition to the companies' proposal—a direct request for a loan of $10,000,000 was sent to the Department of State by 'Abd al-'Aziz through diplomatic channels on June 26, 1941; *FR(1941)*, III, 638, n.23a.

[94] Memorandum dated August 8, 1941, of a conversation of Lloyd Hamilton and Fred Davies with Warren Pierson, who had recently discussed the matter with Jones; U.S. Congress, *Petroleum Arrangements with Saudi Arabia*, p. 25445.

summer,[95] and the British did increase and continue the subsidy.[96] Since by that time Britain was receiving a substantial amount of direct Lend Lease, the aid to Saudi Arabia was viewed in Washington as indirect support from the United States, and the companies tried hard to convince 'Abd al-'Aziz that they deserved much of the credit for having engineered this.[97] Except for corporate concern that the British might receive too much credit, and 'Abd al-'Aziz' continuing desire for larger subsidies, all parties appear to have been satisfied by this resolution of the problem.

*

WHATEVER else may be said about the unsuccessful attempt to persuade the American government to aid 'Abd al-'Aziz directly in 1941, it clearly illustrates the fact Saudi oil was *not* a matter of strategic concern to the United States prior to American entry into World War II. Nowhere in the considerable body of surviving documentation is the strategic argument advanced. This is understandable in view of the fact that in 1941 the United States was still a net exporter of petroleum,[98] there was no widespread concern about depletion of American reserves, and the extent of the Saudi fields was not yet fully appreciated. All of this began to change in 1943, and when it did the essential elements of the problem to be solved were already well established. The vast reserves of Saudi Arabia were under the personal control of an

[95] Memorandum of conversation, Hull with British Ambassador Viscount Halifax, May 7, 1941, *FR(1941)*, III, 632; memorandum, Jones to Hull, August 6, 1941, *FR(1941)*, III, 642.

[96] The British subsidy to 'Abd al-'Aziz was $12,090,000 in 1942 and $16,-618,250 in 1943; for the rest of the war, aid was provided jointly by the United States and Great Britain. The companies reluctantly continued their cash advances, but decreased them from $2,433,222 in 1941 to $2,307,023 in 1942, and $79,651 in 1943, when direct American Lend Lease made this unnecessary; U.S. Congress, *Petroleum Arrangements with Saudi Arabia*, p. 25381.

[97] Rodgers testimony, ibid., pp. 24828-29; cable, Davies (San Francisco) to Lebkicher (Jidda), October 28, 1941, ibid., p. 25430.

[98] The United States was a net exporter of crude and products combined in every year from 1923 through 1947; it became a net importer in every year from 1948 on; *Petroleum Facts and Figures*, 1971, pp. 283-87.

astute Wahhabite Bedouin Arab monarch with a critical need for money to ensure internal political stability, an abiding antipathy for the world Zionist movement, and a deep suspicion of the British and the Hashimite rulers of Iraq and Transjordan. An American company, Standard Oil of California, had established a promising concession there, and had brought in another American concern, The Texas Company, to help market Bahrain oil. As the extent of the Saudi reserves became known, both companies were concerned that inability to market this additional oil might seriously jeopardize the concession, and Texaco especially was concerned about the amount of additional investment that now appeared to be required if they attempted to go it alone. Jersey and Socony-Vacuum had the markets and were crude-short, but they had been blocked from coming to terms with Socal and Texaco by Gulbenkian and the French under the Red Line Agreement. And overshadowing all of this was uncertainty as to the future role of the British, long the dominant Western power in the Middle East, deeply entrenched in Iranian and Iraqi oil, and with a long record of close association with Shell and Anglo-Iranian in matters of mutual interest. As will be seen, it took the American government and the American companies seven years to work their way through this labyrinth and establish a firm linkage between the United States and Saudi Arabian oil.

II

STRATEGIC PLANNERS

SIGNIFICANT shifts in national power relationships seldom occur instantaneously. In retrospect, their beginning can be traced far back in time, and considerable concern over the implications frequently can be found among those closest to and responsible for the area at the time. This early interest often goes unnoticed by the general public because of more pressing contemporary issues, and when the potential problem matures into a real one there is a great hue and cry to find those responsible for "letting this happen." In retrospect, however, it will usually be found that there was inadequate time, talent, and resources to deal adequately with all potential problems simultaneously, and what actually was done was the product of competing interest groups—bureaucratic, corporate, and public—rather than careful rational planning. Whether these early participants are viewed as wise or foolish, public-spirited or mean-minded, will depend in large part on the perceived long-term consequences of their decisions, the ideological viewpoint of the observer, and—more prosaically— "whose ox was gored." All of this has been clearly evident in the alarmed discussions that followed the Arab oil embargo of 1973 and the growing awareness of American dependence on foreign oil.[1] For these reasons, it appears useful to go back and place the events of 1943 through 1947 in perspective as background for a discussion of how various individuals and interest groups responded to the initial awareness that the "center of

[1] See, for example, the mildly anticompany U.S. Congress, Senate, Committee on Foreign Relations, Subcommittee on Multinational Corporations, *Multinational Corporations and United States Foreign Policy*, pts. 7 and 8; John M. Blair's bitterly anticompany, *The Control of Oil*; Neil H. Jacoby's procompany, *Multinational Oil: A Study in Industrial Dynamics*; and Anthony Sampson's journalistic, *The Seven Sisters: The Great Oil Companies and the World They Made*.

gravity of world oil production [was] shifting from the Gulf-Caribbean area to the Middle East.''[2]

In the context of this study, the most significant event of the period was the transition of the United States from the position of net exporter to one of net importer of petroleum. The petroleum industry had originated in the United States in the nineteenth century, and for years America was by far the leading producer and consumer, with a comfortable margin for export. But American and world demand continued to grow in the twentieth century, and production followed suit, with an increasingly larger proportion outside the United States. (Comparative production figures for 1938 and 1950 are given in Table II-1.) Domestic American production could not keep up with domestic consumption, and the result was a shift from net exportation to net importation during the 1940s, as shown in Table II-2. Domestic consumption in 1950 stood at 2.4 billion barrels per year, with net imports only 8.4 percent of that figure. But the trend had been firmly established, and by the time of the embargo in 1973 American annual domestic consumption stood at 6.3 billion barrels, with net imports at 34.8 percent.[3] A premonition of this changing situation had run through the American government as early as 1941 when Secretary of the Interior Harold L. Ickes called to the attention of President Franklin D. Roosevelt a steady decline in the ratio of proven domestic reserves to annual production. The ratio had dwindled from an indicated twenty-two-years supply in 1930 to an indicated fifteen-years supply in 1940, and Ickes strongly recommended that ''thought . . . be taken concerning the future of the United States

[2] "Preliminary Report of the Technical Oil Mission to the Middle East," prepared by E. L. DeGolyer for the president and directors of the Petroleum Reserves Corporation, February 1, 1944; pp. 43-44 of bound volume, "Petroleum Reserves Corporation," Records Regarding Claim Against the PRC by Arabian American Oil Company, in Record Group 234, Records of the Reconstruction Finance Corporation (hereafter cited as RG234).

[3] Adapted from U.S. Department of Energy, *Monthly Energy Review*, January, 1979 (Springfield, Virginia: National Technical Information Service, 1979), p. 26. For excellent discussion of this trend and its broader significance, see Joel Darmstadter and Hans H. Landsberg, "The Economic Background," in Raymond Vernon, ed., *The Oil Crisis*, pp. 15-37.

with respect to oil.''[4] An October study that he forwarded to Roosevelt went further; it argued that "the United States must have extra-territorial reserves to guard against the day when our steadily increasing demand can no longer be met by our domestic supply.''[5] The problem was thus clearly foreseen as early as October 1941, two months before the attack on Pearl Harbor, and concern for future petroleum resources was woven all through the discussions of America's strategic position during the war years.

Although it was quite logical for Roosevelt's secretary of the interior to be the first to alert the president to this problem, much of Ickes' subsequent role was due to his personality rather than his position. An attorney, Progressive Republican, municipal reformer, and conservationist, Ickes was appointed to the cabinet

TABLE II-1
WORLD CRUDE OIL PRODUCTION, 1938 AND 1950
(Thousands of metric tons and percent of world total)

Area	1938		1950	
United States	164,107	(60.0)	266,704	(51.1)
Other Western Hemisphere	44,303	(16.2)	106,036	(20.3)
Total Western Hemisphere	208,410	(76.2)	372,745	(71.4)
Saudi Arabia	67	(−)	26,617	(5.1)
Iran	10,359	(3.8)	32,259	(6.2)
Iraq	4,363	(1.6)	6,650	(1.3)
Other Middle East	1,133	(0.4)	20,450	(3.9)
Total Middle East	15,922	(5.8)	85,976	(16.5)
Other Eastern Hemisphere[1]	49,084	(18.0)	63,361	(12.1)
Total Eastern Hemisphere	65,006	(23.8)	149,337	(28.6)
Total World	273,416	(100.0)	522,082	(100.0)

SOURCE: See Table I-2.
1. Combined "Communist" and "Noncommunist" from the original data.

[4] Letter, Ickes to Roosevelt, December 1, 1941, folder: "Correspondence with the President," box 666, item 11, Record Group 253, Records of the Petroleum Administrator for War (hereafter cited as RG253).
[5] Memorandum by Ralph K. Davies, deputy petroleum administrator, October 15, 1941, attached to letter, Ickes to Roosevelt, October 18, 1941, folder: "Correspondence with the President," box 666, item 11, RG253.

TABLE II-2

UNITED STATES EXPORTS AND IMPORTS OF CRUDE PETROLEUM AND PRODUCTS,
1938-1950

(Thousands of 42-gallon barrels)

Year	Exports	Imports	Net Exports	Domestic Consumption	Net Exports as a Percent of Consumption
1938	194,145	54,308	139,837	1,136,705	12.3
1939	189,549	59,060	130,489	1,230,486	10.6
1940	130,940	83,751	47,189	1,326,620	3.6
1941	108,830	97,142	11,688	1,485,779	0.8
1942	116,907	35,966	80,941	1,449,908	5.6
1943	149,957	63,412	86,545	1,521,426	5.7
1944	207,616	92,311	115,305	1,671,263	6.9
1945	182,983	113,619	69,364	1,772,685	3.9
1946	153,123	137,676	15,447	1,792,786	0.9
1947	164,477	159,389	5,088	1,989,803	0.3
1948	134,674	188,144	− 53,470	2,113,678	− 2.5
1949	119,376	235,559	− 116,183	2,118,250	− 5.5
1950	111,306	310,261	− 198,955	2,375,057	− 8.4

SOURCE: Adapted from *Petroleum Facts and Figures, 1971*, pp. 284-85.

post in 1933 and became one of the president's most ardent and colorful supporters. Sixty-seven years old in 1941, he was full of energy and engaged in controversy throughout his entire time in office. Of particular importance to this study was his easy access to the president, his skill in bureaucratic in-fighting, and his long-standing feud with Secretary of State Cordell Hull. As will be seen, all of these characteristics influenced the course of the debate over oil policy that ensued.

In 1941 Ickes warned of future domestic shortages, but where should the United States turn for "extraterritorial reserves"? Increasing awareness of the extent of Middle Eastern resources in general and Saudi Arabian resources in particular pointed clearly toward that area. A study prepared by the vice chief of naval operations for the Joint Chiefs of Staff in May 1943, gave the proven reserves of the United States as 20 billion barrels, and by comparison estimated Middle Eastern reserves (unproven but from major structures) as being in excess of 56 billion barrels (Table

II-3). Three decades later these figures seem ridiculously conserv-ative, but the ratio of American to Middle Eastern reserves appears to have been essentially correct.[6] And since within the Middle East the vast undeveloped reserves of Saudi Arabia were already under concession to an American company, what could be more logical than to focus attention on ways and means to ensure con-tinuing American control of that concession? None of this suggests how or against whom the concession should be protected, but it does point out that the incentive to do something about Saudi Arabia was quite strong within the American government from 1943 onward. The basic interest was generated by the changing pattern of resources, and not by any manipulation of the oil com-panies.

One additional point should be made. When one turns to the

TABLE II-3

UNITED STATES NAVY ESTIMATE OF MIDDLE EAST PETROLEUM RESERVES,
MAY 1943

(Billions of 42-gallon barrels)

Country	Proven Reserves[1]	Minimum Additional Estimated Reserves[2]	Minimum Total Estimated Reserves
Saudi Arabia	2.2	20.0	22.2
Iran	7.7	10.0	17.7
Iraq	4.5	10.0	14.5
Kuwait	1.0	1.0	2.0
Total	15.4	41.0	56.4

SOURCE: Letter, Vice Chief of Naval Operations to Joint Chiefs of Staff, May 31, 1943, folder: "CCS463.7(5-31-43)," JCS Decimal Files, 1942-45, Record Group 218, Records of the Joint Chiefs of Staff, National Archives, Washington, D.C.

1. Given as "Known Reserves" in the original.
2. Given as "Estimated Additional Reserves from known major structures, in excess of . . . (amount)" in the original.

[6] U.S. Congress, Senate, Committee on Interior and Insular Affairs, *Geopolitics of Energy*, by Melvin A. Conant and Fern R. Gold, p. 33. As of January 1975, total recoverable reserves (including reserves yet to be discovered) were estimated at 242 billion barrels for the United States and 663 billion barrels for the Middle East.

question of *who* the Saudi concession should be protected *against*, it is surprising to find that the perceived enemy was Great Britain and the British-controlled oil companies. Nowhere in the accessible British archives is there any evidence of a British plan in the 1940s to actually displace the American concessionaire, and the fears appear to have been a product of past British activity, the close ties between the British government and the British companies, intense commercial rivalry, and 'Abd al-'Aziz playing one power off against the other to achieve his own ends.

As mentioned earlier, Britain had a long-term imperial strategic interest in protecting the Middle East against hostile interests, and this had led to subsidies to most of the Persian Gulf sheikhdoms in exchange for pacts requiring British approval of actions involving foreign governments and companies. With the British government a majority shareowner in Anglo-Iranian, there was considerable intertwining of corporate and government interests by zealous oil men and civil servants. A specific example often cited in the 1940s was Kuwait, where Gulf Oil had been able to obtain a concession only on agreement that it be shared fifty-fifty with Anglo-Iranian, and on further agreement that Gulf would not sell Kuwait oil in Anglo-Iranian's marketing territory.[7] In Saudi Arabia, Britain's wartime interest was in sustaining the political stability and pro-Ally pan-Arab influence of 'Abd al-'Aziz,[8] but Socal and The Texas Company were constantly worried that too strong a British connection might somehow weaken the American hold on the yet-to-be-developed concession.[9]

'Abd al-'Aziz was fully aware of all of this, and archival records in Washington and London suggest that he subtly fanned those fears to increase his chances for financial support. He told the

[7] Diary entry for September 18, 1943, Harold L. Ickes MSS (hereafter cited as Ickes MSS).

[8] "Annual Report on Saudi Arabia for 1943," in letter, S. R. Jordan (Jidda) to Anthony Eden, February 28, 1944, vol. 40283, file E1293/1293/25, FO/371, British Foreign Office Political Correspondence (hereafter cited as V[volume number, file number]), FO/371.

[9] Memorandum of W.S.S. Rodgers, April 27, 1944, U.S. Congress, Senate, Special Committee Investigating the National Defense Program, *Petroleum Arrangements with Saudi Arabia, Hearings*, pt. 41, p. 25385.

British minister that he "preferred Britain to guide the destinies of the Arab States rather than America . . . [because of Britain's] . . . long record of co-operation with and friendship for the Arabs,"[10] and that he was "cautious towards the Americans . . . because . . . the shadow of Zionism loom[ed] behind all their activities in the Middle East."[11] But to Brigadier General Patrick Hurley, who paid him an official visit in June 1943, the king expressed "great faith and confidence in the United States," and concern "that the British government . . . intend[ed] to force imperialistic rule on the Arab states." He acknowledged that "his country for its own safety and welfare needed the friendship and assistance of a strong foreign power" but preferred American "development of his natural resources" because American companies were not "dominated by an imperialistic government" that would keep them from being "subject to Saudi Arabian authority." He was worried, however, about general "British . . . economic penetration in the Middle East."[12] Such sentiments were temporarily reassuring, but the concession was essentially a personal grant from an absolute monarch, 'Abd al-'Aziz was in his sixties, there was no assurance as to the attitude of his successor, and such thoughts did nothing to allay the long-term fears of the American companies and the American government.[13]

[10] "Annual Report on Saudi Arabia for 1943," in letter, Jordan to Eden, February 28, 1944, V40283, E1293/1293/25, FO/371.

[11] "Annual Report on Saudi Arabia for 1944," in letter, Jordan to Eden, February 9, 1945, V45546, E952/952/25, FO/371. It should be noted that these sentiments might be more indicative of what the British minister wanted to hear than what the king actually said.

[12] Letter, Hurley to Roosevelt, June 9, 1943, folder: "Saudi Arabia," box 86, President's Secretary's File, Franklin D. Roosevelt MSS (hereafter cited as Roosevelt MSS). Foreign service officers and Aramco officials who were in Saudi Arabia during the 1940s recall that 'Abd al-'Aziz expressed himself very much along the lines reported in Hurley's letter; interviews with Ambassador James S. Moose, Jr., in Washington, Kentucky, April 24, 1977, Floyd W. Ohliger, in Pineville, Pennsylvania, July 21, 1977, and Thomas C. Barger, in La Jolla, California, August 26, 1977. Ambassador Moose was of the opinion that the king gradually shifted his focus to the United States as it became apparent that an American connection could be of more help to him financially.

[13] Letter, Alexander Kirk (American minister in Cairo) to Secretary of State,

The general view from Washington in the early 1940s was thus concern over the adequacy of future American reserves, growing awareness of the extent of Saudi reserves, and fear that the American concession there might somehow drift into British control. As will be seen, renewed overtures by W.S.S. Rodgers and H.C. Collier for governmental aid to Saudi Arabia in early 1943 set in motion a policy debate that quickly polarized around three basic positions. The first was for direct participation in the Saudi concession by the American government itself, along the lines of British government participation in Anglo-Iranian. This was the route initially favored by Secretary Ickes, President Roosevelt, and Undersecretary of the Navy William C. Bullitt, and it took shape in the abortive Petroleum Reserves Corporation. Vehement opposition came from all of the nonparticipating American oil companies. The second was for a negotiated agreement with the British to ensure rational development of Middle Eastern oil and equitable treatment for all parties. This was the avenue favored by Secretary of State Hull and initially supported by most of the major American oil companies, but ultimately opposed by the Texas-based independents. The third was to leave the matter essentially in the hands of private enterprise, with diplomatic backing to ensure "equal access" to foreign fields and protection for American investments overseas. The leading proponents of this view in 1943 were John A. Brown, president of Socony-Vacuum, and Ralph W. Gallagher, president, and Wallace E. Pratt, vice-president, of Standard Oil (New Jersey). The fate of these three proposals will be discussed in the balance of this chapter (the British model), in Chapter III (the Anglo-American agreement), and in Chapter V (the private initiative).

*

IT MIGHT be well to emphasize the fact that governmental concern over the adequacy of American reserves predated the Rodgers and

February 28, 1942, enclosing a memorandum dated February 26, 1942, prepared by Roy Lebkicher (Casoc representative in Jidda) for transmission to Frederick A. Davies (Casoc president, in San Francisco), 890F.6363 Standard Oil Company/ 138, RG59.

Collier initiative by at least eighteen months. In July of 1941, on the recommendation of Economic Advisor Herbert Feis, the Department of State acquired the services of Max W. Thornburg, then vice-president of Bahrain Petroleum, as Petroleum Advisor, to coordinate the work of the geographic divisions on this subject.[14] In his initial survey of the field, especially the continuing negotiations with Mexico over their expropriation of American companies in 1938, Thornburg noted the decline in domestic American reserves and called attention to the future importance of Western Hemisphere reserves for American security.[15] This concern was transmitted to Ralph K. Davies, former vice-president of Standard Oil of California, who in June had been appointed by Harold Ickes as his deputy petroleum coordinator for national defense (later petroleum administrator for war).[16] After a detailed study of the reserve situation, Davies concluded that there was, indeed, a long-term problem, and that from his viewpoint "the petroleum resources of Mexico, Colombia, Venezuela, and other Caribbean countries must be considered to be reserves for the United States."[17] As mentioned earlier, Davies' study was trans-

[14] Manuscript, "The Petroleum Division," October 1944, p. 1, folder: "Petroleum-General," box 48, Harley Notter files, RG59; and excerpts from memorandum, Thornburg to Charles Rayner (his successor as petroleum advisor), January 8, 1944, folder: "Oil," Duffield Office Files, RG80. The exact title and reporting status of the petroleum advisor went through a number of changes in the 1940s, but for simplicity these nuances are omitted in this study. Although Feis recommended Thornburg for this position, the original suggestion that State hire someone to perform this function came from Ickes; John W. Frey and H. Chandler Ide, *A History of the Petroleum Administration for War, 1941-1945,* p. 20.

[15] Memorandum, Robert E. Allen to Davies, September 29, 1941, folder: "Correspondence with the President," box 666, item 11, RG253.

[16] Frey and Ide, p. 14. For an excellent account of the origins of the Petroleum Administration for War, see Michael B. Stoff, *Oil, War and American Security: The Search for a Natural Policy on Foreign Oil, 1941-1947,* pp. 10-33.

[17] Memorandum, Davies to Ickes, October 15, 1941, folder: "Correspondence with the President," box 666, item 11, RG253. The roles of Thornburg and Davies later came under sharp scrutiny when it became known that both continued to draw enough salary from their former employers to make up the difference between their former and their government salary. Davies studiously avoided official handling of any matters that might involve a conflict of interest and came

mitted by Ickes to Roosevelt in October 1941, and it indicates that the initial discussions on American reserves arose in connection with Mexican rather than Saudi oil.[18]

In 1943 this Western Hemisphere concern produced another result that was to have considerable indirect consequences seven years later in the Middle East. Since 1937 Venezuela had been attempting to renegotiate its concessions with Jersey, Gulf, and Shell to raise its revenue above the 7 to 10 percent royalties in the original contracts.[19] When heavy wartime tanker losses forced

through the war with his reputation intact. Thornburg was not as careful, and—guilty or not—was forced into resignation in July, 1943, on charges of conflict of interest. It is difficult to tell from the available documents whether the charges were valid or were the product of bureaucratic infighting, but it is clear that Ickes was the source of charges against Thornburg that Roosevelt passed on to Sumner Welles; Cordell Hull, *The Memoirs of Cordell Hull*, 2, p. 1517; Herbert Feis, *Seen from E.A.: Three International Episodes*, p. 120; memorandum, Ickes to Roosevelt, March 15, 1943, enclosing a photostat of a letter, Thornburg to Welles, February 8, 1943, and memorandum, Roosevelt to Welles, June 30, 1943, all in folder: "State Department 6-12/43," box 90, President's Secretary's file, Roosevelt MSS; manuscript, "The Petroleum Division," p. 8, folder: "Petroleum-General," box 48, Harley Notter files, RG59; and U.S. Congress, *Petroleum Arrangements with Saudi Arabia*, p. 24857. Neither Thornburg nor Davies were central figures in the petroleum reserves episode, but Thornburg was among the original supporters of a government corporation to purchase the oil offered by Rodgers and Collier (memorandum, Thornburg to Hull, March 27, 1943, 800.6363/1141-1/2, RG59), and Davies privately advised Ickes against attempting to buy all of the stock in Casoc and against going through with the pipeline agreement he had negotiated with Socal, Texas, and Gulf, because of mounting opposition from other companies (diary entries for August 22, 1943, and February 6, 1944, Ickes MSS). Additional circumstantial evidence pointing to Davies' basic loyalty to the government included Davies' chagrin over the fact that John D. Rockefeller, Jr. had supported Collier rather than Davies for the presidency of Socal just before the war (diary entry for December 19, 1943, Ickes MSS), and the fact that Davies organized the American Independent Oil Company (Aminol) to take up a concession in the Kuwait Neutral Zone rather than return to Socal after the war; Sanger, pp. 165-68, and Longrigg, pp. 214-15.

[18] This early concern was reinforced by a special study conducted by the Petroleum Industry War Council in January, 1942, which reported that "the reserves of this country may soon be inadequate"; Frey and Ide, p. 174.

[19] This account of the Venezuelan negotiations in 1943 and 1944 is based on the excellent analyses in Stephen G. Rabe, "Energy for War: United States Oil Diplomacy in Latin America During World War II," in David H. White, ed.,

a 35 percent reduction in Venezuelan production in 1942, the decline in government revenue prompted a renewed attack on the concession issue. Undersecretary of State Sumner Welles took the position that corporate reasonableness was in the American national interest, and Wallace Pratt convinced the Jersey board that the time had come for a change. Pratt went to Caracas in December 1942, along with a Gulf representative, to negotiate with Attorney General Gustavo Manrique Pacanins and Herbert Hoover, Jr., who had been retained to advise the Venezuelan government. Also present was Max Thornburg, invited by President Isais Medina Angarita as an "unofficial" observer. In fact, the negotiations consisted of drafting a new Venezuela petroleum law, working from an outline provided by Thornburg. Negotiations were complete by February 1943, and the law was enacted by the Venezuelan Congress on March 13. When it became operative, its complex royalty and taxing provisions had the effect of dividing profits fifty-fifty between the companies and the government and increasing Venezuela's oil revenue by 80 percent. Under the Internal Revenue Act of 1918, the tax portion of these payments to Venezuela were deductible from corporate taxes due to the American government.[20] Seven years later this law provided the precedent for the "fifty-fifty" agreement with 'Abd al-'Aziz.

In exact parallel with the Venezuelan negotiations, the Department of State had begun to take increasing interest in Saudi Arabia. A permanent legation and consulate, with James S. Moose, Jr., as chargé, was established at Jiddah on April 22, 1942 (with the rank raised to minister resident in July 1943); an agricultural mission under Karl Twitchell was sent out in the summer; wartime air transit rights were negotiated in August;[21]

Proceedings of the Conference on War and Diplomacy, 1976, pp. 125-132; and Henrietta M. Larson, Evelyn H. Knowlton, and Charles S. Popple, New Horizons, 1927-1950, pp. 479-84.

[20] For an analytic history of the Internal Revenue Act of 1918, see U.S. Congress, Senate, Committee on Foreign Relations, Subcommittee on Multinational Corporations, Multinational Corporations and United States Foreign Policy, pt. 8, pp. 351-56.

[21] U.S. Department of State, Foreign Relations of the United States (1942), IV, 559-75 (hereafter cited as FR[date]). For an excellent account of this developing interest, see Miller, pp. 50-60.

and by December, the Division of Near Eastern Affairs had initiated action to make Saudi Arabia eligible for Lend Lease. In a most significant memorandum dated December 14, 1942 Division Chief Paul H. Alling argued that the country was critical not only for wartime transit rights and petroleum supplies, but also for its long-term strategic value. "Dhahran," he said, "gives every promise of being one of the world's most important oilfields," and considering the forecasts of limited American reserves, retention of the Saudi concession was "an American interest of the highest importance." He noted that "the British Government . . . [had] . . . been supplying [extensive] material assistance," and there was "a definite possibility that the British . . . [would] . . . require a *quid pro quo* [in petroleum rights] at the end of the war,"[22] a development that would be inimicable to the national interest. The recommendation was forwarded with a strong endorsement by Dean Acheson as acting secretary of state to Lend Lease Administrator Edward R. Stettinius, Jr., who transmitted it to the president on January 11, 1943.[23] A recommendation that Lend Lease aid be extended to Saudi Arabia was thus on Roosevelt's desk by the early part of January 1943, backed by State Department concern that failure to match British support might ultimately weaken the American hold on a strategically valuable concession.

Meanwhile, events in Saudi Arabi had taken an interesting turn. The only legal currencies at the time were English gold sovereigns and Saudi silver riyals, and wartime hoarding of gold and silver was taking so much money out of circulation that an increasing portion of the British subsidy had to be in sovereigns and newly minted riyals in order for the government to meet its cash requirements. To correct this problem, the British treasury in 1942 proposed a paper riyal, backed by reserves provided by the British

[22] Memorandum, Alling to Assistant Secretaries of State Adolph A. Berle and Dean Acheson, December 14, 1942, 890F.24/20, RG59.

[23] Letter, Acheson to Stettinius, January 9, 1943; and letter, Stettinius to Acheson, January 12, 1943, *FR(1943)*, IV, 854-55. In his letter to Stettinius, Acheson deleted Alling's long-term strategic argument and confined the rationale to short-term wartime needs.

government, and controlled (to avoid inflation) by a currency board composed of the Saudi minister in London, a representative of the British treasury and a representative of the British government.[24] The proposal was eventually dropped for lack of confidence that paper riyals would be accepted in the Bedouin culture,[25] but its discussion in the winter of 1942 did nothing to allay the concern of Socal and The Texas Company over increasing British influence. To Rodgers and Collier, the Currency Board appeared to be an opening wedge, and they decided that the time had come to press again for direct American aid to counteract this trend.

In early February 1943, the two company presidents came to Washington and discussed the Saudi situation with Harold Ickes in his capacity as Petroleum Administrator for War, and then with as many others as possible, including Undersecretary of State Sumner Welles, Undersecretary of War Robert P. Patterson, Assistant Secretary of the Navy William C. Bullitt, Executive Officer of the Army-Navy Petroleum Board Captain (later Admiral) Andrew F. Carter, and State Department Economic Advisor Herbert Feis, Political Advisor Wallace S. Murray, and Near Eastern Division Chief Paul H. Alling.[26] Their basic argument was detailed in a memorandum Rodgers sent to Ickes on February 8. The Saudi reserves, he said, "are of the same order of magnitude as the great reserves of Iran and Iraq," and their importance to the United States "becomes increasingly apparent as . . . [America faces] . . . a decrease in our national reserves and an increase in national consumption." The British have "made substantial advances to the Saudi Arabian government in cash and supplies . . . , made possible by the aid which the British government has been receiving from the United States . . . [and they are] . . .

[24] Letter, Kirk to Secretary of State, February 28, 1942, 890F.6363 Standard Oil Company/138, RG59; telegram, Kirk to Secretary of State, January 18, 1943, *FR(1943)*, IV, 856-57; telegram, Shullaw (Jidda) to Secretary of State, March 22, 1943, *FR(1943)*, IV, 864; and memorandum: "General Considerations in Connection with Economic or Financial Assistance to Saudi Arabia," April 3, 1943, 890F.51/51-1/4, RG59.

[25] Telegram, Kirk to Secretary of State, May 11, 1943, *FR(1943)*, IV, 869.

[26] U.S. Congress, *Petroleum Arrangements with Saudi Arabia*, p. 24854.

now insisting that Saudi Arabian financial requirements be met
by an internal note issue . . . controlled by the British Currency
Control Board.'' He expressed deep concern ''over the rapidly
increasing British economic influence in Saudi Arabia because of
the bearing it . . . [might] . . . have on the continuation of purely
American enterprise there after the war.''[27]

As a countermeasure, Rodgers proposed ''direct aid from the
United States . . . instead of indirect aid through the British.''
Specifically, he suggested that ''the British . . . creditor position
. . . be transferred to the United States by crediting to the British
government's Lend Lease debt to the United States the amount
now owed by the Saudi Arab government to the British government.'' If this were done, Casoc would ''be willing to set aside
as a reserve for the United States an amount of oil having an
equivalent value . . . [to] . . . be made available to the United
States government at prices well under world prices.'' Additional
Lend Lease aid could be backed by additional reserves ''under
a similar arrangement.'' In addition, Rodgers offered the government an option contract for crude or products ''at an agreed upon
percentage of the going price in . . . world markets . . . at the
time the . . . government elected to take delivery.''[28] In short,
Rodgers proposed direct Lend Lease aid to Saudi Arabia and
offered a petroleum reserve to the United States as a quid pro
quo.

On Februray 16, Ickes discussed the Saudi matter with Roosevelt over lunch, but his diary and his subsequent congressional
testimony are clear that the discussion dealt only with the general
importance of Saudi reserves, and not with Lend Lease.[29] Two
days later, however, Roosevelt signed the Lend Lease order for

[27] Memorandum attached to letter, Rodgers to Ickes, February 8, 1943, folder:
"Middle East," box 666, item 11, RG253. The memorandum is reproduced in
U.S. Congress, *Petroleum Arrangements with Saudi Arabia*, pp. 25385-87.
[28] Ibid.
[29] Diary entry for February 20, 1943 (recording a luncheon meeting with Roosevelt on February 16), Ickes MSS; U.S. Congress, *Petroleum Arrangements with
Saudi Arabia*, pp. 25233-36.

Saudi Arabia that Stettinius had sent him a month earlier.[30] It is possible that Roosevelt signed the order with greater enthusiasm as a result of his conversation with Ickes, but the documentary evidence is clear that the recommendation on which he acted originated with the Department of State and not with the two oil companies.[31]

The Rodgers initiative may not have been responsible for Lend

[30] Letter, Roosevelt to Stettinius, February 18, 1943, *FR(1943)*, V, 859.

[31] This interpretation differs from most previous accounts, which at least infer that Rodgers and Collier were the prime movers behind Lend Lease for Saudi Arabia. The error can be traced to a 1948 Senate investigation already cited as U.S. Congress, *Petroleum Arrangements with Saudi Arabia*. The report of that investigation (U.S. Congress, Senate, Special Committee Investigating the National Defense Program, *Navy Purchases of Middle East Oil*), written without benefit of State Department or Lend Lease Administration records, placed Ickes' discussion with Roosevelt on February 16 in juxtaposition to the Lend Lease order on February 18 and incorrectly inferred a causal relationship. Most subsequent accounts appear to have been based on this investigation and report, and they have repeated the inference that Rodgers and Collier were the prime movers; see Benjamin Shwadran, *The Middle East, Oil, and the Great Powers*, p. 309; George W. Stocking, *Middle East Oil: A Study in Political and Economic Controversy*, pp. 96-98; Gerald D. Nash, *United States Oil Policy, 1890-1964: Business and Government in Twentieth Century America*, p. 171; Sampson, p. 95; Gabriel Kolko, *The Politics of War: The World and United States Foreign Policy, 1943-1945*, pp. 295-96; and Louis Turner, *Oil Companies in the International System*, pp. 40-41. An exception is the previously cited doctoral dissertation by Joseph W. Walt, "Saudi Arabia and the Americans, 1928-1951" (pp. 206-220), which used State Department records and which also notes the prior State Department recommendation but does not conclude which was most influential. Stoff (pp. 58-61) notes the earlier State Department recommendation but still gives prime credit to the oil companies. Hull, in his *Memoirs* (2, p. 1512) credits the State Department and does not even mention the oil companies' initiative. Aaron D. Miller (*Search for Security: Saudi Arabian Oil and American Foreign Policy, 1939-1949*, pp. 65-71) gives a full description of the corporate overture but correctly gives principal credit to the Department of State. Further evidence that the State and corporate initiatives were entirely separate is provided by the fact that a document dated San Francisco, December 29, 1942, and presumably given to State by Socal at a later date argues for a naval petroleum reserve in Saudi Arabia and does not even mention Lend Lease; memorandum: "The Importance of Foreign Oil Reserves in General and Saudi Arabian Reserves in Particular," December 29, 1942, folder: "Saudi Arabian-Oil," box 6, Records of the Petroleum Division, RG59.

Lease for Saudi Arabia, but it brought the Saudi issue dramatically
to the attention of key people all over Washington. In his lunch
with Roosevelt on February 16, Ickes told the president that "the
Government ought to have a financial interest in . . . [the] . . .
American oil concessions in Arabia and that there would probably
never be a better time to do it than now,"[32] but he took no
immediate action to cause that to happen. Instead, Rodgers' proposal for a contractual petroleum reserve was taken up by the
special Committee on International Petroleum Policy that had been
organized inside the State Department in mid-January under the
chairmanship of Economic Advisor Herbert Feis.[33] The department's Division of Commercial Policy and Agreements favored
solving the problem by international agreement, but Feis overrode
those objections in favor of more direct action. In late March he
obtained Hull's endorsement of a proposal to establish a Petroleum
Reserves Corporation for the purpose of negotiating option-contracts with Casoc and other American companies holding reserves
outside the United States.[34] On March 31 Hull forwarded the
proposal for comment to the Departments of the Interior, War,
and the Navy, where it met a storm of protest.[35] Ickes, Knox, and
Stimson favored what they saw as the British model—direct purchase of a controlling interest in Casoc to warn all comers that
the concession was a vital American interest. The ensuing bureaucratic debate bogged down in the face of more pressing war-

[32] Diary entry for February 20, 1943, Ickes MSS.

[33] Manuscript, "The Petroleum Division," pp. 11-12, folder "Petroleum-General," box 48, Harley Notter files, RG59.

[34] Ibid.; memorandum, Harry C. Hawkins (Division of Commercial Policy and
Agreements) to Hull, March 23, 1943, folder: "Jan.-Mar. 1943," box 4, Leo
Pasvolsky files, RG59; Hull, *Memoirs*, 2, p. 1517; memorandum (Feis) to Hull,
March 22, 1943, folder: "Petroleum Reserves Corporation Organization, Feb. 22-
July 2, 1943," Records of the Petroleum Division, RG59. This proposal was
further explained in a memorandum from Thornburg to Acheson and four others
on March 27, 1943, folder: "Jan.-Mar. 1943," box 4, Leo Pasvolsky files, RG59.
Thornburg wanted to use the terms of these option-contracts to control the behavior
of the oil companies vis-à-vis other countries.

[35] Hull, *Memoirs*, II, 1518.

time matters and might never have resurfaced had it not been revived by another chain of events.[36]

For some time, the Army-Navy Petroleum Board (ANPB) had been studying long-term petroleum supplies for combat operations, and in late April it concluded that a serious problem existed.[37] Forecasts were for a worldwide shortage in 1944, and to remedy this, the board recommended to the Joint Chiefs of Staff increased or new refinery capacity in Colombia, Venezuela, Iran, Palestine, and Saudi Arabia.[38] Because of the considerable investment of steel that would be required, the Joint Chiefs directed a detailed interagency study prior to reaching a decision.[39] In the midst of this discussion, Assistant Secretary of the Navy Bullitt took up the cause of direct action to protect the reserves in Saudi Arabia. Using the Army-Navy Petroleum Board recommendation as a springboard, Bullitt called on Ickes, Knox, and Stimson, warning that an increase in "our supply from petroleum outside the United States has become . . . essential for efficient prosecution of the war,"[40] and that "the British are already laying plans to get a cut in . . . [the] . . . Saudi Arabian field."[41] Specifically, he proposed that a Petroleum Reserves Corporation be set up immediately and charged with acquiring a controlling interest in Casoc.[42]

[36] Feis, *Seen from E.A.*, pp. 110-19.

[37] Army-Navy Petroleum Board 21/1, April 20, 1943, and minutes of Army-Navy Petroleum Board meeting of April 21, 1943, both in folder: "CCS463.7 (4-9-43) Sect. 1," Joint Chiefs of Staff Decimal Files, 1942-47, RG218.

[38] Enclosure "C" to JCS 281/1, May 26, 1943, in folder: "CCS463.7 (4-9-43) Sect. 1," JCS Decimal Files, 1942-47, RG218.

[39] Enclosure "A" to JCS 281/1, May 26, 1943, in folder: "CCS463.7 (4-9-43) Sect. 1," JCS Decimal Files, 1942-47, RG218.

[40] Draft letter from Bullitt to Roosevelt, June, 1943, in U.S. Congress, Senate, Committee on Foreign Relations, Subcommittee on Multinational Corporations, *A Documentary History of the Petroleum Reserves Corporation 1943-1944*, p. 3.

[41] Diary entry for May 29, 1943, Ickes MSS. See also entries for June 12 and 13.

[42] Diary entry for June 12, 1943, Ickes MSS, and draft letter from Bullitt to Roosevelt, June, 1943, in U.S. Congress, *Documentary History of the Petroleum Reserves Corporation*, pp. 3-6.

Apparently concluding that matters were moving too slowly, and that the Department of State would block a decision indefinitely, Bullitt broke the logjam by getting the issue directly to Roosevelt through the Joint Chiefs of Staff. At Bullitt's instigation, the Vice Chief of Naval Operations, Admiral F. J. Horne, proposed that the Joint Chiefs recommend "a government corporation to acquire . . . a controlling interest in the great Saudi Arabian petroleum reserves."[43] With misleading statements that "the Secretary of State approves the idea in principle," and that "the British have already approached the King of Saudi Arabia with a view to acquiring rights," Horne obtained the concurrence of the Joint Chiefs,[44] and the recommendation was made to Roosevelt by Admiral William D. Leahy on June 8.[45] Unfortunately, Leahy was poorly informed on the subject, and the letter of recommendation lacked clarity. As a result, Roosevelt decided on what he apparently thought was a simpler course of action. He asked Leahy to suggest to Hull that the American minister to Saudi Arabia and Captain Carter negotiate a new concession there as a "naval fuel oil reserve similar to those . . . in the United States."[46] He was apparently unaware that Casoc already had virtually all of the promising areas under concession. When this misguided presidential directive descended on official Washington, it provoked a flurry of heated activity.

In addition to reaching Roosevelt through the Joint Chiefs, Bullitt had stirred Knox and Stimson to action, and on the same day that the Joint Chiefs met, those two called on Ickes to press the proposal for a petroleum reserves corporation.[47] Two days

---

[43] Letter, Horne to Joint Chiefs of Staff, Ser.069139 (SC)JJ7/EG, May 31, 1943, folder: "CCS463.7 (5-31-43)," JCS Decimal File, RG218.

[44] Minutes of Joint Chiefs of Staff meeting of June 8, 1943, folder; "CCS463.7 (5-31-430)," JCS Decimal File, RG218.

[45] Letter, Leahy to Roosevelt, June 8, 1943, *FR(1943)*, IV, 921.

[46] Memorandum, Leahy to Hull, June 11, 1943, *FR(1943)*, IV, 921-22. Leahy was quite chagrined when he discovered that he had inadvertently misled the president; minutes of Joint Chiefs of Staff meeting of June 15, 1943, folder: "CCS463.7 (5-15-43)," JCS Decimal File, 1942-45, RG218.

[47] Diary entry for June 13, 1943, Ickes MSS; memorandum, Feis to Hull, June 10, 1943, 890F.6363/80, RG59.

later (June 10), Ickes sent a formal recommendation to Roosevelt to that effect, arguing that "by the end of 1944 we shall be unable to produce sufficient crude oil to meet the petroleum requirements of the armed services and . . . civilian economy," a situation that should be remedied by "immediate action to acquire a proprietory interest in foreign petroleum reserves," beginning with Saudi Arabia. His point was that American involvement of "a sovereign character" was required to counteract British competition of an equally "sovereign character."[48] Although Ickes' proposal was thoroughly consistent with his own prior thinking, its submission appears to have been the result of Bullitt's prodding, and it was also on the table when a meeting was called on June 11 by Director of War Mobilization James F. Byrnes to untangle the matter.

The meeting in Byrnes' office was attended by Ickes, Knox, Stimson, Feis, and Brigadier General Boykin C. Wright of the War Department, and it took up three alternatives.[49] Roosevelt's suggestion that a mission be sent to negotiate a new concession was obviously based on a misapprehension that would have to be corrected. Feis continued to hold out for his option-contract plan, as one that had already been proposed by Casoc and one that would be much less objectionable to competing companies. He also insisted on a State Department veto over any overseas actions taken by the proposed Petroleum Reserves Corporation. Arrayed against him in favor of buying a controlling interest in Casoc were Secretaries Knox and Ickes, and when Hull subsequently declined more than token backing for Feis,[50] the outcome was foreordained. A series of four subsequent meetings of lesser officials[51] under the chairmanship of Undersecretary of War Patterson worked out

[48] Letter, Ickes to Roosevelt, June 10, 1943, in U.S. Congress, *Petroleum Arrangements with Saudi Arabia*, pp. 25237-38.

[49] Memorandum, Wright to Stimson, June 11, 1943, U.S. Congress, *Documentary History of the Petroleum Reserves Corporation*, pp. 6-8; memorandum, Feis to Hull, June 11, 1943, 890F.6363, RG59; and Hull to Roosevelt, June 14, 1943, *FR(1943)*, IV, 922-24.

[50] Feis, *Seen from E.A.*, pp. 110-21.

[51] Patterson, Wright, Colonel W.E.R. Covell, Feis, Bullitt, Carter, and Undersecretary of the Interior Abe Fortas; *FR(1943)*, IV, 930.

a recommendation to the president for the collective endorsement of the secretaries of state, the interior, war, and the navy.

The document that went to the president was a curious one, and the cards were stacked against Feis. All parties recommended establishment of a petroleum reserves corporation to acquire an interest in Saudi Arabian reserves, but Hull and Feis believed that the document left to Roosevelt the choice between the option-contract method and the purchase of a controlling interest in Casoc.[52] A careful reading of the document, however, makes it clear that it was written in such a way that the four departments were unanimously recommending purchase of a controlling interest, and appending the pros and cons of the two methods as an indication of the positions that had been considered before a recommendation was made.[53] For Roosevelt to choose the option-contract method he would have had to read the document very carefully and then act against the unanimous recommendation of four of his cabinet members. But according to Feis, the final discussion with the president was "jovial, brief, and far from thorough." The president approved the proposal with "a boyish note of enjoyment."[54]

The option-contract plan brushed aside in so cavalier a fashion was a simplified version of the one proposed by Rodgers in Feb-

[52] Hull, II, 1520, and Feis, *Seen from E.A.*, p. 121.

[53] Letter, Hull, Stimson, Forrestal, and Ickes to Roosevelt, June 26, 1943, *FR(1943)*, IV, 924-30. What appears to be an earlier draft of this document (minus reference to the option-contract) appears in U.S. Congress, *Documentary History of the Petroleum Reserves Corporation*, pp. 9-12.

[54] Feis, *Seen from E.A.*, p. 122. The PRC was formally established by the Reconstruction Finance Corporation on June 30 to meet the deadline of July 1 when legislative authority for such action was due to expire. But Roosevelt's note directing formal incorporation said, "No further action of any kind is to be taken pending further instructions from me." Memorandum, F.D.R. to the secretary of commerce, June 29, 1943, folder: "Petroleum Reserves Corporation-Organization," box 1, WAC and WAA Fiscal Records, RG234. The project itself was approved by Roosevelt in a meeting with Feis and Wallace Murray the first week in July; Feis, *Seen from E.A.*, p. 122; diary entry for July 3, 1943, Ickes MSS; memorandum, Feis to Hull, July 3, 1943, 800F.6363/52, RG59; and letter, Hull to Roosevelt, July 6, 1943, U.S. Congress, *Documentary History of the Petroleum Reserves Corporation*, pp. 12-13.

ruary. Under it the Petroleum Reserves Corporation would have taken over responsibility for payment of advance royalties through Casoc to 'Abd al-'Aziz. In return Casoc would have set aside a percentage of its estimated reserves for delivery to the American government at a favorable price when needed. Instead, the plan approved by Roosevelt called for purchase of "100% of the stock" of Casoc from Socal and The Texas Company, by reimbursing those two companies for all expenses to date and allotting them a percentage of future production. Socal and The Texas Company would be given preference for a contract to manage the Saudi operation, but "the right to manage and operate the several structures could be granted to the oil companies making the best competitive bids."[55] The board of directors of the Petroleum Reserves Corporation was to consist of the secretaries of state, the interior, war, and the navy, and State retained the right to veto "any major projects or undertakings" and to conduct all "major negotiations with foreign governments."[56] The audacity of the overall plan was possibly reflective of the mood of wartime Washington.

Thus was born the Petroleum Reserves Corporation and the attempt on the part of the American government to buy a controlling interest in the California Arabian Standard Oil Company. In retrospect, it was the product of an exaggerated concern that America was running out of oil and a desire to ensure access to foreign reserves. The idea that the government should establish an interest in Saudi reserves to counter British influence came from Socal and Texaco and was nurtured all spring in the Department of State's option-contract plan. Ickes preferred emulating British tactics by buying a controlling interest in Casoc, but he took little action to cause that to happen and the project bogged down in bureaucratic debate. It was resurrected by excessively pessimistic military forecasts and a burst of energy out of the navy that rode roughshod over the Department of State's caution and led to adoption of the bolder course. Nowhere in the accessible record is there evidence of advance consultation with the oil in-

[55] Letter, Hull, Stimson, Forrestal, and Ickes to Roosevelt, June 26, 1943, *FR(1943)*, IV, 929.
[56] Ibid.

dustry on this plan, and nowhere is there indication that anyone other than Feis gave serious consideration to the opposition it might evoke.

*

MOST ACCOUNTS of the negotiations that followed between the government and the owners of Casoc report that Socal and The Texas Company simply refused to sell even a minority interest.[57] That version, widely accepted for years, was based on Ickes' subsequent testimony before a congressional committee, the memoirs of Herbert Feis, and published records of the Petroleum Reserves Corporation.[58] But a careful examination of newly accessible material in Ickes' confidential diary, the original records of the Petroleum Reserves Corporation, and unpublished State Department documents, reveals quite another story. There is strong evidence that Casoc and The Texas Company had almost completed an agreement to sell a one-third interest in return for government financing of the Ras Tanura refinery when Ickes himself terminated the negotiations because of pressure brought to bear by Socony-Vacuum and Jersey. Ickes simply omitted mention of the role of the other two companies in his congressional testimony, Feis was not privy to this part of the story, and the published minutes of a key meeting of the PRC board differ significantly from the version in the original PRC records. The sequence of events is fairly complex and requires careful reconstruction.

[57] See, for example: Hull, II, 1520-21; Raymond F. Mikesell and Hollis B. Chenery, *Arabian Oil: America's Stake in the Middle East*, pp. 91-92; Walt, pp. 244-46; Nash, p. 173; Stocking, pp. 98-99; Shwadran, pp. 322-23; Mira Wilkins, *The Maturing of Multinational Enterprise: American Business Abroad from 1914 to 1970*, pp. 277-78; Sampson, pp. 96-97; Robert B. Krueger, *The United States and International Oil: A Report for the Federal Energy Administration on U.S. Firms and Government Policy*, pp. 50-51; Stephen D. Krasner, *Defending the National Interest: Raw Materials Investments and U.S. Foreign Policy*, pp. 190-97; Stoff, pp. 84-86; and Miller, pp. 81-82.

[58] U.S. Congress, *Petroleum Arrangements with Saudi Arabia*, pp. 25240-43; U.S. Congress, *Navy Purchases of Middle East Oil*, pp. 13-14; Herbert Feis, *Petroleum and American Foreign Policy*, pp. 38-39; Feis, *Seen from E.A.*, pp. 131-33; U.S. Congress, *Documentary History of the Petroleum Reserves Corporation*, pp. 40-41.

Ironically, responsibility for getting the PRC started fell to its staunch opponent, Herbert Feis. On his initiative, further lower-level discussions were held, a detailed plan was sent by Hull to Roosevelt, and on July 15 Roosevelt met with Hull, Feis, Ickes, Knox, and Patterson to complete agreement on a course of action.[59] After some argument by Ickes to keep Secretary of Commerce Jesse H. Jones off the PRC board because of his Texas oil connections, it was agreed that the board would consist of the secretaries of state, war, the navy, and the interior, and that the legal home of the PRC would be transferred from the Reconstruction Finance Corporation (under Jesse Jones) to the Office of Economic Warfare (headed by Leo T. Crowley). Ickes would be president of the new organization, and Roosevelt asked that Ickes and Feis conduct the negotiations with Socal and The Texas Company.[60] Without waiting for even the first meeting of the PRC board, Ickes called back into federal service Texas lawyer and former Undersecretary of Interior Alvin C. Wirtz to handle the detailed negotiations,[61] and notified Collier and Rodgers that he wanted to meet with them in Washington on August 2.

At this point another chain of events intersected these plans. It will be recalled that interest in Saudi reserves had been rekindled by an Army-Navy Petroleum Board analysis of world-wide refining capacity in preparation for military operations from 1944 onward. That planning had continued during the debates over the PRC, and by late July the Army-Navy Petroleum Board had decided to recommend to the Joint Chiefs that a refinery of about 100,000 barrels/day capacity be constructed in Saudi Arabia at an estimated cost of $100,000,000, "based on anticipation of military supply needs in the Southwest Pacific, . . . [and on]

[59] Memorandum, Feis to Hull, July 6, 1943, 890F.6363/53, RG59; letter, Hull to Roosevelt, July 6, 1943, *FR(1943)*, IV, 930-32; diary entry for July 20, 1943, Ickes MSS.

[60] Diary entry for July 20, 1943, Ickes MSS; Feis, *Seen from E.A.*, pp. 122-26; minutes of meeting of directors of the Petroleum Reserves Corporation, August 9, 1943, bound volume: "Petroleum Reserves Corporation," Records Regarding Claim Against PRC by Arabian American Oil Company, RG234.

[61] Diary entry for August 8, 1943, Ickes MSS; U.S. Congress, *Documentary History of the Petroleum Reserves Corporation*, pp. 14-16.

. . . the conclusion that it would be advisable in the long run to draw on Persian Gulf supplies for the quantities needed, rather than on our own domestic supplies."[62] Preliminary plans for such a refinery had been completed by the foreign division of the Petroleum Administration for War,[63] and Commodore Carter of the Army-Navy Petroleum Board met with Feis and then with Ickes to suggest that "an undertaking of this size . . . would . . . require Government financing and . . . [would] . . . be certain to influence the Company's attitude in the prospective discussions with this Government about stock purchase."[64] Ickes liked the idea. He immediately requested that the Army-Navy Petroleum Board assign the project to the PRC, and indicated that he would link financing for the refinery with the stock-purchase proposal in his discussions with Collier and Rodgers.[65]

The first round of negotiations took place on August 2 and 3 in Washington. Representing the government were Ickes, Wirtz, Feis, and Mortimer Kline, general counsel for the PRC. Representing the companies were Collier and H. H. MacCarrigal of Socal, and Rodgers, Vice-President Clarence E. Olmsted and General Counsel Harry T. Klein of The Texas Company. When Ickes announced that the purpose of the negotiations was purchase of "all of the common stock of the California Arabian Standard Oil Company" he found himself "looking into the faces of two surprised and shocked individuals." He went on to note that "the

[62] Memorandum, Feis to Hull, July 26, 1943, *FR(1943)*, IV, 933-34.

[63] Letter, Ickes to Carter, August 6, 1943, 890F.6363/63, RG59. The director of the Foreign Division of the PAW until November, 1943, was James Terry Duce, in temporary government service from his normal duties as vice-president of Casoc. The record suggests, however, that the preliminary designs for the Saudi refinery were developed by his associate director C. Stribling Snodgrass.

[64] Memorandum, Feis to Hull, July 26, 1943, *FR(1943)*, IV, 933-34.

[65] Ibid.; diary entry for August 1, 1943, Ickes MSS; letter, Ickes to Carter, August 6, 1943, 890F.6363/63, RG59; minutes of board meeting of August 9, 1943, U.S. Congress, *Documentary History of the Petroleum Reserves Corporation*, pp. 16-20. The Joint Chiefs approved the Saudi refinery project on August 3 and the ANPB assigned the project to the PRC on August 10; letter, Carter to Ickes, August 10, 1943, folder: "CCS463.7 (4-9-43) Sect. 2," JCS Decimal File, 1942-45, RG218.

Army and Navy want to build a 100-million-dollar refinery in Saudi Arabia . . . [and] . . . if this refinery was to be built with United States funds it would have a bearing upon and a connection with the agreement that we were trying to formulate."[66] Neither Collier nor Rodgers had any interest whatsoever in selling the entire Saudi operation to the American government,[67] but the refinery would require a very heavy additional investment, and Rodgers was not opposed to exploring the possibility of a deal. After an initial two days of unproductive fencing, both sides retired to take stock of their positions. Most of the subsequent two and a half months of negotiations were carried on between Collier, Rodgers, and Wirtz, with Feis participating down to the last two weeks, when he incurred Ickes' disfavor and was shunted aside.

It was clear from the outset that the companies would not surrender 100 percent ownership, and Ralph Davies persuaded Ickes to reduce the demand to 51 percent, in line with the British precedent in Anglo-Iranian.[68] Ickes obtained Roosevelt's concurrence for this on August 29, but when Wirtz presented it to Collier and Rodgers, they were still adamantly opposed. They did, however, indicate a willingness to consider sale of a one-third interest, and negotiations proceeded on that basis.[69] Both sides now began to use pressure tactics to obtain the best possible deal. Ickes decided that the government's need for a refinery in Saudi Arabia gave the companies too much leverage, so he moved to create alternatives. On September 10 he called in Colonel J. Frank Drake, president of Gulf Oil, and broached the possibility of sale of an interest in the Kuwait concession in return for diplomatic efforts to break the restrictive sales agreement with Anglo-Ira-

[66] Diary entry for August 8, 1948, Ickes MSS.
[67] Follis interview; Long interview; Rogers' testimony, U.S. Congress, *Petroleum Arrangements with Saudi Arabia*, p. 24845.
[68] Diary entries for August 22 and 29, 1943, Ickes MSS.
[69] Memorandum, Feis to Hull, September 10, 1943, 890F6363/67, RG59; memorandum, Feis to Hull, September 15, 1943, 890F.6363/65, RG59; minutes of meeting of PRC board, September 13, 1943, U.S. Congress, *Documentary History of Petroleum Reserves Corporation*, pp. 26-28.

nian.[70] Drake was sufficiently interested for Ickes to be able to keep informal discussions going with Gulf during the balance of his talks with Collier and Rodgers. In parallel with the Gulf initiative, Carter came up with the idea of alternative refinery sites as leverage against the companies and arranged for the Army-Navy Petroleum Board to recommend to the Joint Chiefs that consideration be given to Kuwait and Bombay. Those sites were eventually rejected by the Joint Chiefs on grounds that a Saudi refinery was more consistent with American interests, but they were also under discussion during the balance of the negotiations with Collier and Rodgers.[71] Pressure tactics were not restricted

[70] Memorandum, Feis to Hull, September 9, 1943, 890F.6363/66, RG59; diary entry for September 12, 1943, Ickes MSS; minutes of meeting of PRC board, September 13, 1943, U.S. Congress, *Documentary History of the Petroleum Reserves Corporation*, pp. 26-28; memorandum, Feis to Hull, September 16, 1943, 890F.6363/69, RG59; memorandum, Feis to Hull, September 16, 1943, 890G.6363/418, RG59; and diary entry for September 18, 1943, Ickes MSS.

[71] Minutes of PRC board meeting, September 13, 1943, U.S. Congress, *Documentary History of Petroleum Reserves Corporation*, pp. 26-28; minutes of meeting of Joint Chiefs of Staff, November 9, 1943, minutes of meeting of Army-Navy Petroleum Board, November 30, 1943, letter, Hull to Leahy, December 15, 1943, and report by the Army-Navy Petroleum Board to the Joint Chiefs of Staff, January 25, 1944, all in folder: "JCS 463.7 (4-9-43)," JCS Decimal File, 1942-45, RG218. The Kuwait refinery would have been operated by Gulf, and the one at Bombay by the Standard Vacuum Oil Company, a jointly owned subsidiary of Jersey and Socony-Vacuum. The Joint Chiefs' final decision was in deference to Hull's argument that "inadequate development of American-owned oil concessions in the Middle East would endanger the continuance of those concessions in American hands, . . . and therefore . . . refining facilities in that area should be so planned as to use oil produced by American companies in the Middle East and, where possible, should be located in the country of production." Letter, Hull to Leahy, December 15, 1943, *FR(1943)*, IV, 950-52. See also, memorandum, Sappington to Alling, November 11, 1943, 800.6363/1375, RG59. During the course of the negotiations, Ickes had issued a "letter of intent" to Casoc authorizing expenditure of funds in advance of a contract. When negotiations broke down, this letter of intent was canceled, and after the war Aramco filed a claim against the government for unnecessary expenses thereby incurred; letter, Davies (Casoc) to Ickes, September 6, 1943; memorandum for the president of the Petroleum Reserves Corporation, November 3, 1943, and letter, Davies (Casoc) to Ickes, November 4, 1943, all in folder: "Refinery File," box 1, Records Regarding Claim Against PRC by Arabian American Oil Company, RG234.

to the government, however. Twice The Texas Company summarily broke off negotiations, and twice Ickes coaxed them back to the bargaining table, once with a sizable concession and once with stern talk.[72]

A concrete proposal was ultimately made by the companies themselves, and by October 14 agreement had been reached on all but one clause. In return for a one-third interest in Casoc, the government would pay $40,000,000. The government would have a preemptive right to purchase up to 51 percent of Casoc's production in peacetime and 100 percent in wartime, and would have the right to block sales to third parties that it considered contrary to American national interests. Except for those restrictions, Casoc would continue to operate as a normal commercial enteprise. The government would advance the funds for the refinery and be repaid from future earnings, minus a yet-to-be-agreed-upon amount to be written off as "war costs" because of capacity in excess of peacetime needs.[73] This last item was the sticking point. A careful reading of the record reveals that Wirtz was incorrect in his assumption that "war costs" had been covered when the figure of $40,000,000 was agreed upon, and Rodgers was correct in his subsequent contention that "war costs" were to be in addition to the $40,000,000. But that is getting ahead of the story. Suffice it to say that by October 14 this point was the only issue remaining to be resolved.

All parties to these negotiations had been bound to absolute secrecy, but the petroleum industry is not noted for its ability to keep projects of this magnitude quiet; by early October it was widely known that the government was attempting to buy an interest in the Saudi operation. Opposition from other segments of the industry was swift and vehement. Since the beginning of

---

[72] Diary entries for October 10 and 18, 1943, Ickes MSS.

[73] Memorandum, Feis to Hull, September 25, 1943, 890F.6363/70, RG59; minutes of meeting of the PRC board, September 28, 1943, memorandum, Brigadier General W.E.R. Covell to Stimson, September 28, 1943, and minutes of the meeting of PRC board, October 14, 1943, U.S.Congress, *Documentary History of the Petroleum Reserves Corporation*, pp. 29-36; diary entry for October 18, 1943, Ickes MSS.

1943 Ralph W. Gallagher, president of Standard Oil (New Jersey), and John A. Brown, president of Socony-Vacuum, had been urging the State Department to define a foreign oil policy for the United States,[74] and in June, Gallagher had gone one step further. In a letter to Undersecretary of State Sumner Welles, he urged consideration of a long policy statement drafted by Wallace Pratt, calling for worldwide development of petroleum resources by private enterprise, protection of private enterprise by enforceable commercial treaties, and collaboration with other governments ". . . to insure the orderly development of . . . [those] . . . resources."[75] When word spread that the government was about to move in a very different direction from this, Jersey and Socony-Vacuum rose in opposition.

Formal public opposition centered in the Foreign Operations Committee of the Petroleum Administration for War (PAW), chaired by Jersey vice-president Orville Harden. The Foreign Operations Committee included senior representatives of all major companies operating overseas, including Jersey, Socony-Vacuum, Socal, The Texas Company, Gulf, Sinclair, Atlantic, Cities Service, Tidewater, and Union Oil, and its task was oversight of the overseas petroleum war effort. Intense discussions by that group in late October produced by November 5 a strongly worded document titled "A Foreign Oil Policy for the United States," concurred in by all members except Socal, Texas, Gulf, and Sinclair (Sinclair's representative had been out of the country).[76] The document went well beyond Pratt's memorandum in favor of private development of petroleum reserves, strong diplomatic support for American companies abroad, and some type of international machinery to ensure "efficient and orderly development

[74] Memorandum, Adolph A. Berle, Jr. to Hull, Welles, Feis, and Thornburg, January 6, 1943, 800.6363/1125, RG59. The overtures by Gallagher and Brown appear to have been the catalysts that prompted formation of Feis' intradepartmental petroleum committee in mid-January.

[75] Letter, Gallagher to Welles, June 14, 1943, 800.6363/1209, RG59.

[76] Frey and Ide, pp. 65-66, 276, 330, 391-93; roster of Petroleum Industry War Council, December 8, 1941; folder "Petroleum Industry War Council," box 687, item 15, RG253; and memorandum, Rayner to Stettinius, December 27, 1943, 800.6363/1404-1/2, RG59.

of the world's oil resources.'' It added clear and specific disapproval of direct government participation in the oil business in any way, shape, or form.[77] This was formidable opposition, coming as it did from a bloc of companies that up to that point had contributed mightily to Ickes' success as Petroleum Administrator for War. Ickes' rather formidable power within the bureaucracy rested heavily on his close relationship with Roosevelt, which in turn was based on his proven ability to get results. If this corporate revolt were to undermine cooperation with the PAW itself, the consequences would be serious. Ickes had great courage and skill as a bureaucratic infighter, but part of his success lay in knowing when to fight and when to deftly change direction.

The personal risk of continuing down this course had already been brought to Ickes' attention by Harold Wilkinson, British Petroleum representative in the United States, in a private conversation in early September. Although he was reacting only to rumor, Wilkinson expressed great concern about the effect that American governmental entry into the oil business would have on Anglo-American relations, and told Ickes that ''he was making a first class personal 'boner' '' by risking sacrificing the ''general acclaim'' he had won by keeping ''his early pledge that the government should not interfere with industry.''[78] This argument appears to have had had little effect in September, but clear opposition by members of the Foreign Operations Committee in late October created quite another situation.

The denouement began on October 15. In what was to have been the final round of negotiations, Rodgers took the position that ''war costs'' should be set at 40 percent of the total refinery costs, which would create an additional $40,000,000 of expense

[77] ''A Foreign Oil Policy for the United States,'' November 5, 1943, 800.6363/1406-1/2, RG59.

[78] Letter, Wilkinson to Sir William B. Brown (Petroleum Division, Ministry of Fuel and Power, London), September 2, 1943, V34210, A9193/3410/45, FO/371. Wilkinson went on to remind Ickes ''that Secretaries come and go'' and asked him ''what he would think of the matter if certain of his present colleagues whose names I forebore to mention but whose defects he had so often discussed with me, were given the post [of PRC chairman], which of course might happen at any time.'' Ickes reportedly replied, ''That is a very terrible thought.''

for the government. To those who knew Rodgers well, this was a typical negotiating stance, but Wirtz was furious. On the morning of October 15, Ickes called in Wirtz and Undersecretary of the Interior Abe Fortas to discuss the situation. It became evident that Wirtz had let himself be trapped by not clearly including the "war cost" factor in the original offer of $40,000,000 for a one-third interest, and Ickes suggested calling negotiations off "without any statement of my reason or saying what my intentions were as to the future."[79] There was no prior consultation with the PRC board or with the president, or with anyone else, and the record suggests that on the morning of October 15, Ickes was simply considering a negotiating tactic borrowed from Rodgers' own book in order to regain control of the situation. At any rate, Rodgers came in at 10:20 a.m., he and Ickes exchanged strong words about whether or not the companies had mentioned "war costs" in the earlier negotiations (Feis' records show that they had), and at Ickes' request, Rodgers retired to come back later in the day with a revised proposal.[80]

One other appointment followed immediately after Rodgers' that morning. With uncannily propitious timing, Ickes' next visitor, at 11:30 a.m., was John Brown of Socony-Vacuum, who came in escorted by Ralph Davies to discuss rumors he had heard of the Casoc negotiations. Brown minced no words. He had learned of the negotiations from his "vice president in London," and "his company and others in the foreign field didn't like the idea of government competition." Direct governmental participation in the oil business would have a dangerously disruptive effect around the world, with no real improvement in the government's already existing ability to draw on corporate reserves anywhere in times of emergency. And "to go through with the deal would mean stirring up . . . wide-spread political criticism" in the United States. Brown "talked very quietly," and Ickes

---

[79] Diary entry for October 18, 1943, Ickes MSS. Rodgers took a similar extreme position at the outset of the final negotiations with Jersey and Socony-Vacuum in 1946.

[80] Ibid.

noted in his diary that he "could not but acknowledge . . . that there was a good deal to what he said."[81]

The Socony intervention proved to be decisive. When Rodgers sent up a new draft proposal that afternoon, Ickes took only a cursory look and sent Wirtz down to adjourn the conference, "*sine die.*" Rodgers was unaware of Brown's visit, and he later testified that "I have done a lot of trading in my day, but I never had anything like that happen before . . . at 11 o'clock in the morning, I thought we were together on the thing. At 2 o'clock in the afternoon, without any preliminaries at all, we were told that the deal was off."[82] Any doubts Ickes may have had about resuming negotiations were resolved during the next two weeks, during which time it became clear that a majority of the Foreign Operations Committee shared Brown's views. On November 3, the matter came up for final discussion in a meeting of the PRC board. According to Ickes' confidential diary, and according to the original draft of the minutes of that meeting, Ickes discussed "the reasons that had led to the discontinuance of our conversations" including the "talk with Brown of Socony-Vacuum,"[83] but obtained the board's agreement that he had acted properly in breaking off negotiations because "The Texas Company was unwilling to reach an agreement . . . except on terms . . . which were deemed to be unreasonable."[84] The final minutes of that meeting were altered to omit all reference to Brown's intervention,[85] and Ickes' subsequent testimony placed full responsibility

[81] Ibid.

[82] U.S. Congress, *Petroleum Arrangements with Saudi Arabia*, pp. 48468 and 24870.

[83] Diary entry for November 7, 1943, Ickes MSS; see also minutes of PRC board meeting of November 3, 1943, bound volume: "Petroleum Reserves Corporation," Records Regarding Claim Against PRC by Arabian American Oil Company, RG234.

[84] Minutes of PRC Board meeting of November 3, 1943, U.S. Congress, *Documentary History of Petroleum Reserves Corporation*, p. 40.

[85] The original draft of those minutes contain the following paragraph: "The Board was unanimously of the opinion that the interests of the people of the United States and its foreign oil industry required the assistance of the United States

for termination of the negotiations on Rodgers' unreasonable-ness.[86] He obviously preferred not to discuss his yielding to quiet behind-the-scenes presssure. To conclude matters, however, that same PRC board meeting voted to discontinue negotiations with Casoc for government funding of the Ras Tanura refinery. When the Foreign Operations Committee published its opposition to Ickes' plan two days later, the negotiations for a government interest in Casoc were already a dead issue and Ickes was well on the way to shifting the blame to the companies themselves.

One footnote should be added to all of this. Both Carter and Davies reported that the probable reason for the opposition from Jersey and Socony-Vacuum was the fact that they were "nego-tiating for a third interest [in Casoc], which, of course would not be obtainable if the Government took a third interest."[87] Nowhere in the accessible record is there evidence that such negotiations were actually taking place in 1943, and industry rumors to that effect were probably based on the fact that negotiations had taken place before the war, and Jersey and Socony-Vacuum were known to have a long-standing interest in gaining access to the Saudi

---

Government or an agency thereof in the protection of American controlled oil reserves. *Mr. Ickes reported a conversation held with Mr. John Brown, President of Socony-Vacuum Oil Company and a discussion ensued as to whether the in-terests of the American nationals and groups in the Middle East could be safe-guarded successfully in the absence of governmental participation. After a full discussion, the negotiator was authorized by the Board to proceed with the con-tinuation of the negotiations"* (emphasis added); minutes of PRC board meeting of November 3, 1943 (unsigned), bound volume; "Petroleum Reserves Corpo-ration," Records Regarding Claim Against PRC by Arabian American Oil Com-pany, RG234. The published version of those minutes delete that portion in italics above and substitute the following: *"The Directors expressed deep regret that the CASOC representatives had been unable or unwilling to appreciate the urgency of and need for the assistance of this Government"* (emphasis added); minutes of PRC board meeting of November 3, 1943 (signed by Abe Fortas as Secretary), U.S. Congress, *Documentary History of Petroleum Reserves Corporation*, p. 41. It appears that no actual vote was taken on termination of the negotiations, and Ickes simply "corrected" the minutes to conform to his understanding of the discussion before they were distributed.

[86] U.S. Congress, *Petroleum Arrangements with Saudi Arabia*, pp. 25240-43.
[87] Diary entry for November 7, 1943, Ickes MSS.

reserves. Whatever the specific reason for the objections raised by Jersey and Socony-Vacuum, it is clear that their opposition to government participation in Casoc was decisive.

An entirely separate question is whether or not the PRC attempt to buy into Casoc was a sound idea in the first place. It probably was not. The stated intent was to establish a direct American governmental interest in the Saudi concession in order to protect it from British encroachment, and the concept was borrowed directly from what was perceived to be the British governmental role in Anglo-Persian from 1914 onward. But 1943 was not 1914, and the United States was not Great Britain. In the context in which the attempt was made, it was an inappropriate measure based on a rather simplistic understanding of the issues involved. In retrospect, it is difficult to see what the American government could have accomplished toward protection of the concession as a part-owner that it could not have accomplished without such status. And the furor that the attempt generated made subsequent attempts to reach a national consensus on foreign oil policy exceedingly difficult. But that is getting ahead of the story again; there was more to come.

III

DIPLOMATS

THE IDEA of negotiating an understanding with the British on Middle Eastern oil had been gestating within the Department of State for well over a year before it was advanced as a serious alternative to Ickes' stock purchase plan.[1] As will be seen, the diplomatic solution was based on a carefully thought-out assessment of the problem that might have carried the day had it not run afoul of bureaucratic infighting over who would control American oil policy. The core of the proposal was a binational governmental advisory commission to oversee production and marketing of Middle Eastern oil in order to ensure equity to all parties, including the producing countries. Such a commission, properly constituted and controlled, would have been consistent with the long-range interests of companies already established in the Middle East, and companies such as Jersey and Socony-Vacuum initially supported the idea. Texas-based independents and others not established in the Middle East feared worldwide regulation of production and imports to their disadvantage, and they were basically opposed. Given quiet negotiation and careful attention to the interests of all parties (including the American consumer), the idea might possibly have worked, but Ickes' continuing impetuosity as president of the Petroleum Reserves Corporation created a firestorm of domestic opposition to "government in

---

[1] In his 1948 testimony Ickes strongly implied that he was the source of the proposal for an Anglo-American accord (U.S. Congress, Senate, Special Committee Investigating the National Defense Program, *Petroleum Arrangements with Saudi Arabia, Hearings*, p. 25243), and at least one subsequent account (Benjamin Shwadran, *The Middle East, Oil, and the Great Powers*, p. 333) incorrectly gave him credit for it. As with much of Ickes' testimony, his phrases were technically correct but highly misleading. This style was characteristic of most of his argumentation during the controversy over the Petroleum Reserves Corporation and the Anglo-American Agreement.

business'' and destroyed any possibility of Senate ratification of
such an agreement. A Texas Railroad Commission was all right
for Texas, but not for the world.[2]

As noted earlier, State Department interest in developing a
coherent foreign oil policy began in July 1941 with the appointment of Max Thornburg as petroleum advisor. As grounding for
his new assignment, Thornburg commissioned a long-term internal study on the history and dynamics of American foreign oil
policy, and by August 1942 he had on his desk a most interesting
interim report.[3] Defining foreign oil policy as "the expression of
the national will in connection with foreign oil operations, or with
foreign trade in oil, through either Government agencies or private
enterprise,'' the report went on to analyze the changing relationship between government and business in each major country in
the world. Among other things, it concluded that the worldwide
trend toward nationalization had diminished "the companies' independent role in foreign oil policy'' and had forced the American
government into a more active stance, with or without the acquiescence of the companies. It further noted that "the cartel
movement . . . [appeared to] . . . bear within it the seeds of
development in the direction either of *a State-controlled engine
of commercial and political policy*, or of commercial internation-
alism'' (emphasis added).[4] In other words, the report suggested

---

[2] Though oddly named for the task it ultimately assumed, the Texas Railroad
Commission was an elective body that held regular hearings and regulated all
Texas fields to keep total production roughly equal to market demand and thereby
ensure both conservation and price stability. The system evolved in the 1930s to
control the burgeoning output of the new East Texas field, and became effective
with passage of the Texas Market Demand Act of 1932, and the federal Connally
("Hot Oil") Act of 1935; Harold F. Williamson et al., *The American Petroleum
Industry*, II, 543-44.

[3] Memorandum: "The Foreign Oil Policy of the United States," August 1,
1942, folder: "Foreign Oil Policy, 1941-42," box 8, Records of the Petroleum
Division, RG59; and memorandum: "The Petroleum Division," by G. M. Richardson Dougall, October, 1944, box 48, Harley Notter files, RG59. The 1942
report is headed "Office of the Petroleum Adviser," but it is not clear whether
it was written by Thornburg or his research assistant, Walton C. Ferris.

[4] Report: "The Foreign Oil Policy of the United States," August 1, 1942,
folder: "Foreign Oil Policy, 1941-42," box 8, Records of the Petroleum Division,
RG59.

the possibility of a government-controlled cartel as a basic instrument of foreign oil policy. This theme ran through much of
the Petroleum Division's thinking for the next two years—including its work on an Anglo-American agreement.

The possibility of Anglo-American talks on Middle Eastern oil
was first broached by Assistant Secretary of State Adolph A.
Berle, Jr. during a visit to London in mid-1942,[5] but the catalyst
for actually doing this came from a variety of corporate initiatives
on both sides of the Atlantic in 1943. Suggestions to the department by Wallace Pratt of Jersey and John Brown of Socony-
Vacuum that State take the initiative in developing a foreign oil
policy prompted formation of the already-mentioned intradepartmental committee under Herbert Feis for that purpose in January
1943.[6] This committee included Thornburg, and it set out to expand on his preparatory work. But when "Star" Rodgers of Texaco and H. C. Collier of Socal made their Saudi reserve proposal
in February, the committee digressed into a debate over whether
the American position could best be protected by an option-contract with Casoc or a "stand-still" agreement with the British.
The option-contract proposal won and went on to become the
Petroleum Reserves Corporation, but Berle and a minority within
the Department of State continued to press for talks with the
British.[7] In June, in the midst of the debate over establishment
of the PRC, Roosevelt asked Hull to undertake a thorough study

[5] Foreign Office Minutes on "America and Oil," October 13, 1943, V34210,
A9286/3410/45, FO/371.

[6] Memorandum by Feis, January 2, 1943, file: "U.S. Foreign Oil Policy thru
1944," box 1, Records of the Petroleum Division; memorandum, Berle to Hull
et al., January 6, 1943, 800.6363/1125; minutes of Petroleum Policy Study Group,
January 11, 1943, 800.6363/1091; and memorandum: "The Petroleum Division,"
by G. M. Richardson Dougall, October, 1944, box 48, Harley Notter files, RG59.

[7] Memorandum, Committee on International Petroleum Policy (Feis) to Hull,
March 22, 1943, folder: "Petroleum Reserves Corporation Organization, Feb. 22-
July 2, 1943," box 1, Records of the Petroleum Division; memorandum, Harry
C. Hawkins (Trade Agreements Division) to Hull, March 23, 1943, folder: "Jan.-
Mar. 1943," Leo Pasvolsky files; memorandum, Berle to Hull, March 24, 1943,
800.6363/1145; RG59. Berle's recommendation of a "stand-still" agreement with
the British specifically included establishment of "an international petroleum
board."

of American petroleum policy, and a new interdepartmental committee for that purpose was established under Feis, with representatives from State, War, the Navy, and the Petroleum Administration for War.[8] Thornburg at this point was in the process of leaving government service, and he was not a member of the reconstituted committee.[9] From the very first, this committee focused on the idea of a negotiated understanding with the British, and by September 28 it had Hull's approval of a draft agreement that it planned to submit to Ickes for his endorsement.[10] The centerpiece of the draft agreement was a binational petroleum board very much along the lines recommended by Thornburg. All of this had gone on in parallel with the PRC discussions with Socal and Texas, and it is clear from what happened later that Ickes had little or no prior knowledge of this project.

Meanwhile, in London, Anglo-Iranian had become concerned about adequate postwar markets for all of the oil being discovered in the Persian Gulf area, and it was worried that uncontrolled competition would lead to price wars and a destabilizing reduction in "the royalties accruing to the Iraki, Persians, Arabs, and the Sheiks of Kuwait and Bahrein." The British company was con-

[8] Minutes of Special Committee on Petroleum, June 15, 1943, folder: "Special Committee on Petroleum," box 28, Records of the Petroleum Division, RG59. The first meeting of the new committee included Feis, Special Assistant to the Secretary of the Navy William Bullitt, Brigadier General Boykin C. Wright (War Department), James Terry Duce (Petroleum Administration for War), Paul Alling (chief, Near East Division), and James Sappington (assistant petroleum advisor under Thornburg), but departmental representation varied at subsequent meetings.

[9] By this time, Ickes' charges against Thornburg had found their way to Roosevelt, Welles, and Hull, and Thornburg was on his way out (see Chapter II, n. 7). Feis reported that the real reason for reconstituting the petroleum policy committee on an interdepartmental basis was to remove the taint of Thornburg's alleged link with Socal; Herbert Feis, *Seen from E.A.: Three International Episodes*, p. 120.

[10] Minutes of Special Committee on Petroleum, July 27, September 21, and September 28, 1943, folder: "Special Committee on Petroleum," box 28; and memorandum, Feis to Hull, September 17, 1943, unmarked black binder, box 3; Records of the Petroleum Division, RG59. A copy of the Draft Agreement, dated September 16, 1943, is enclosed with letter, Wright to Stettinius, December 24, 1943, unmarked black binder, box 3, Records of the Petroleum Division, RG59; a binational petroleum board is the centerpiece of the proposal.

vinced that it was not "possible for the companies themselves to arrive at any agreement," partly because "recent interpretation of the Sherman Anti-Trust Act [precluded the American companies] from having any discussions with one another." Accordingly, it believed that "the British and American governments [should] get together to discuss the entire problem." This should be done soon because it was doubtful whether "after the war a suitable arrangement could be consummated."[11] Anglo-Iranian quietly suggested this to the British government, and the topic was taken up informally in a cabinet meeting on July 14. Churchill, however, preferred not to raise such "far-reaching issues . . . at the present time," and the Foreign Office concluded that it would be better to let the Americans make the first overture.[12] Anglo-Iranian therefore raised the issue on the other side of the Atlantic. On August 9, its New York representative, B. H. Jackson, outlined the suggestion in great detail to James Terry Duce, a Casco vice-president who was then serving as director of the Petroleum Administration for War's Foreign Division.[13] Duce

[11] "Aide-mémoire" of conversation with B. H. Jackson of Anglo-Iranian by James Terry Duce, August 9, 1943; folder: "Middle East," box 666, item 11, RG253.

[12] Foreign Office minutes of July 21, 1943, with attached extract from War Cabinet Conclusion 99/93 of July 14, 1943, V34978, E4582/3710/65, FO/371. The minutes of the War Cabinet meeting do not specify Anglo-Iranian as the company that made the initial suggestion, but the circumstances clearly indicate that this was the company involved. Shell favored discussions between the companies first; Foreign Office minutes of October 1, 1943; V34210, A9286/3410/45, FO/371.

[13] Diary entry for August 15, 1943 (recording a conversation with Duce on August 12), Ickes MSS; memorandum, Duce to Davies, August 12, 1943, folder: "Middle East," box 666, item 11, RG253; and memorandum, Duce to Ickes, August 13, 1943, folder: "Saudi Arabia Pipeline," box 86a, P.S.F., Roosevelt MSS. Duce was an intriguing Washington figure for two decades. Born in England and educated in Colorado as a geologist, he joined The Texas Company in 1918, and became vice-president for government relations for Casoc in 1940. In December, 1941, he temporarily left Casoc to become director of the PAW's Foreign Division, but he returned to Casoc in October, 1943, and served as vice-president for government relations until his retirement in 1959; biographical data provided to the author by Mrs. Ivy O. Duce, letter of August 22, 1977. Although Duce was an extremely competent company representative, he was an independent

fully supported the idea, and gave a complete report of the conversation to Davies and Ickes, who in turn forwarded copies to Roosevelt and Hull.[14] The highest echelons of the American government were therefore fully informed of the Anglo-Iranian viewpoint by mid-August, 1943.

The Duce memorandum had two immediate effects. It provided the Petroleum Division of the Department of State with a more sophisticated rationale for the binational commission that it already favored, and—more importantly—it interested Harold Ickes in negotiating an agreement with the British himself. Shortly after receiving the memorandum Ickes called in the British Petroleum representative in the United States—Harold Wilkinson—to discuss the Petroleum Reserves Corporation, and at the end of the conversation added the statement that he believed "the U.S. and Great Britain should 'get together' on a joint oil policy."[15] Several weeks later he talked with Wilkinson again—this time more specifically. A conference was not only needed, but it should "be so arranged as to fall within the lap of the British and American Oil Agencies rather than under the auspices of the State Department." Since it would be awkward for Ickes himself to issue an invitation, he suggested that he would be happy to see such an overture made from London.[16] Apparently unknown to each other,

thinker and intellectual-in-business who frequently steered his own course. There is no evidence that Casoc itself was in any way connected with the recommendation to Ickes in August, 1943.

[14] Memorandum, Ickes to Roosevelt, August 18, 1943, folder: "Saudi Arabia Pipeline," box 86a, P.S.F., Roosevelt MSS; and letter, Ickes to Hull, August 18, 1943, 800.6363/1281, RG59.

[15] Letter, Wilkinson to Sir William B. Brown (Ministry of Fuel and Power, London), September 2, 1943, V34210, A9286/3410/45, FO/371.

[16] Letter, Wilkinson (then visiting in London) to Brown, September 29, 1943, V34210, A9286/3410/45, FO/371. It has been suggested to the author that Ickes' *real* intent in establishing the Petroleum Reserves Corporation may have been to provide leverage to force the British to negotiate. Although Ickes certainly used this argument later—especially in regard to the pipeline proposal (U.S. Congress, *Petroleum Arrangements with Saudi Arabia*, p. 25244)—there is no evidence in the accessible record that Ickes had that objective at the outset. There is circumstantial evidence that such a view was held by Commodore Carter of the Army-Navy Petroleum Board as early as July 27 (minutes of Special Committee on

two branches of the American government were now laying the groundwork for discussions with the British.

Sometime between September 28 and October 3 Feis took up the State Department's draft agreement with Ickes and informed Ickes that the department planned to send Feis to London shortly for preliminary discussion with the British. The issue as to who was to make American foreign oil policy was suddenly and clearly joined. Ickes moved quickly and with great force. He told Roosevelt that he no longer had confidence in Feis as a conegotiator with Socal and Texas, and when he learned on October 13 that Roosevelt had approved Feis' trip to London he raised such violent objection with Stettinius, the new undersecretary of state, and with his old friend Dean Acheson that on October 14 Stettinius obtained Roosevelt's concurrence to cancel the trip. Shattered by this lack of support in high places, Feis on October 15 tendered his resignation from government service.[17]

The senior survivor of the State Department's petroleum policy planning effort was now Acting Petroleum Advisor James C. Sappington, with support from Political Advisor and former Near Eastern Division Chief Wallace Murray.[18] Taking heart from the

Petroleum, July 27, 1943, folder: "Special Committee on Petroleum," box 28, Records of the Petroleum Division, RG59; this was the only meeting of that committee that Carter attended, and the only meeting in which such views were advanced). But the first time Ickes even hinted at such tactics was in his discussion with Wilkinson in mid-September, and the PRC negotiations with Socal and Texas were already well under way. It is more likely that Ickes shifted his rationale for the PRC to this position as the PRC came more heavily under attack.

[17] Diary entries for October 3 and 18, 1943, Ickes MSS; memorandum, Stettinius to Roosevelt, October 12, 1943, 800.6363/1445, RG59; letter, Feis to Stettinius, October 15, 1943, 111.26/147; letters, Feis to Roosevelt, October 18, 1943, and Roosevelt to Feis, October 21, 1943, folder: "State Department, 6-12/43," box 90, P.S.F., Roosevelt MSS. Feis' resignation occurred on the same day that Ickes broke off negotiations for an interest in Casoc, but there does not appear to have been any direct connection between the two events.

[18] At a meeting shortly after Feis resigned, Stettinius asked Murray to handle "petroleum matters" temporarily, and he and Sappington did so until Charles B. Rayner replaced Sappington as acting petroleum advisor on December 17; memorandum, "The Petroleum Division," by G. M. Richardson Dougall, October, 1944, folder: "Petroleum-General," box 48, Harley Notter files, RG59.

gathering public storm over the Petroleum Reserves Corporation, Sappington and Murray began mobilizing the effort to regain departmental control of the situation.[19] By November 24 Sappington had put together a comprehensive recommendation to Hull that formal discussions be promptly proposed to the British, that approval be obtained from the president for the talks to be controlled by the Department of State, and that appropriate contact be made with the press, other nations, congressional leaders, and (almost as an afterthought) "the American oil companies operating in the Near Eastern area."[20] On December 2, Hull formally proposed talks on Middle Eastern oil to British ambassador Lord Halifax,[21] but the overture to Roosevelt was temporarily delayed. Several days later Sappington learned in a visit from Thornburg that Ickes had told Rodgers that *he* planned to obtain the president's permission to open discussions with the British because the people at State were "a lot of cookie pushers . . . [who] . . . don't even know how to push cookies anymore."[22] At Sappington's instigation, Hull promptly sought and obtained Roosevelt's approval for the American delegation to be headed by a representative of the Department of State.[23] Also as a result of suggestions received from Thornburg in his now private capacity, Sappington recommended that Hull call together an industry panel of the chief executive officers of Jersey, Socony, Socal, Texas, Gulf, Atlantic, and Consolidated Oil to provide advice on the

[19] In addition to his own memorandum, the Petroleum Divisions records show that Sappington drafted most of the documents on this subject signed by Murray and other officers of the Department during November and December, 1943.

[20] Memorandum, Murray to Hull, November 24, 1943, *FR(1943)*, IV, 943-47.

[21] Letter, Hull to Halifax, December 2, 1943, *FR(1943)*, IV, 947.

[22] Memorandum, Alling to Murray et al., December 6, 1943, and Sappington to Murray, December 6, 1943, with attachments, all in unmarked black binder, box 3, Records of the Petroleum Division, RG59. Ickes did discuss the subject with Roosevelt in very general terms in December, but State moved so rapidly that he was on the defensive by the time the issue was really joined over control of the talks; diary entries for December 5, 11, and 26, 1943, Ickes MSS.

[23] Memorandum, Hull to Roosevelt, December 8, 1943, *FR(1943)*, IV, 948. The copy of this memorandum in the records of the Petroleum Division is initialed, "CH OK FDR"; unmarked black binder, box 3, Records of the Petroleum Division, RG59.

negotiations and to make clear to the industry who was in charge.[24]

It is apparent from all of this that the impetus for talks with the British came out of the Petroleum Division, largely in response to concerns over long-term national interests, requests from American and British companies that State take charge, and a challenge from Ickes that if State didn't, he would. The intellectual father of the proposed agreement was Max Thornburg, who continued advising his former colleagues for at least six months after he left government service. By mid-December, then, State was firmly committed to a diplomatic solution and quite prepared to fight it out with Harold Ickes for control of foreign oil policy.

Even more significant, however, was the fact that in the midst of all this State had evolved a basic policy that was to remain constant for over a decade thereafter. On December 1, Sappington pulled all of the loose strands together in a single, tightly argued position paper. The basic need, as he saw it, was for "orderly development of all of the resources of the Middle East, both British and American-controlled, to attain the fullest possible production . . . as soon as possible." From the viewpoint of American security, it was "advisable that Middle Eastern oil be developed to the maximum and that supplies in this hemisphere be . . . conserved" since reserves in the United States and the Caribbean would be far more defensible in time of war. Maximum development was also important for the security of the concessions in general and the Saudi concession in particular because it would assure economic benefit "to the . . . countries which contain the resources," with "proper proportioning of output . . . between countries in the area on the basis of their economic need." The amount of oil in the Persian Gulf was "so great," however, that "the question of distribution . . . [was] . . . at the heart of the problem of full [and equitable] development." In short, Western Hemisphere reserves could best be conserved, and the Saudi concession could best be protected if an adequate market could be carved out for Saudi oil. A working agreement with the

[24] Letter, Thornburg to Sappington, December 8, 1943, and memorandum, Sappington to Murray, December 13, 1943, both in unmarked black binder, box 3, Records of the Petroleum Division, RG59.

British—who had a similar vested interest in markets for Iranian oil—would be highly desirable, but the national interests would also be served by support of company efforts in "the construction and expansion of facilities" to hasten "the over-all development" of the resources.[25] As will be seen in Chapters V and VI, there were more ways than one to accomplish those objectives, but in December 1943, State set out to achieve them through an Anglo-American accord and establishment of a binational petroleum board to oversee the production and marketing of Middle Eastern oil.

There appears to have been no formal consultation with the industry in formulating this initial policy, but it was quite consistent with the November 5 recommendations of the PAW's Foreign Operations Committee. Those recommendations, it will be recalled, had been made by senior executives of virtually all of the American companies operating overseas, with only the companies then in negotiation with Ickes abstaining. Essentially, they called for support of private enterprise, but they went on to advocate the "efficient and orderly development of the world's oil resources," "coordination of national policies" through "voluntary agreements between governments," and an international agency modeled on "the Interstate Oil Compact of 1935" to prevent competitive overproduction.[26] Support from the majors

[25] "Memorandum on the Department's Position . . . ," Sappington, December 1, 1943, folder: "Petroleum Reserves Corporation Activities, 7/3/43-1/1/44," box 1, Records of the Petroleum Division, RG59. The position paper was the basis for extensive discussions within the department on the subject; memorandum, Sappington to Murray, December 1, 1943, 890F.6363/92-1/2, and memoranda, Sappington to Pasvolsky et al., December 4, 1943, and Alling to Pasvolsky, December 9, 1943, folder: "10/43-12/43," box 5, Leo Pasvolsky files, all in RG59. Alling fully supported Sappington's line of reasoning.

[26] Pamphlet, *A Foreign Oil Policy for the United States*, submitted by the Foreign Operations Committee, November 5, 1943, 800.6363/1406-1/2, RG59. The Interstate Oil Compact was a voluntary agreement among the major oil producing states (initially Texas, Oklahoma, Kansas, New Mexico, Colorado, and Illinois) to enact consistent conservation laws, prohibit interstate commerce in "hot oil" produced in violation of those regulations, and establish a fact-finding and informational commission to oversee its operation. It was backed by federal legislation making such a compact legal, and it fell somewhere between laissez-faire and outright government control; Williamson et al., II, 550-51.

could therefore be expected, but neither the department nor the Foreign Operations Committee appear to have given any thought as to the possible reaction of the independents—especially those located in Texas and represented in Congress by Senator Thomas T. (Tom) Connally, chairman of the Senate Foreign Relations Committee. State's initial insensitivity on this point was clearly in evidence in another passage buried in the Sappington memorandum. He argued that "the basic factor in developing widening distribution of Persian Gulf oil . . . [would be] . . . the relative difference in cost of production," with Middle Eastern oil the "cheapest," American oil the most expensive, and Caribbean oil in between. But he noted that if "Middle Eastern oil should enter the United States to meet the postwar need for oil imports, the result should be a further conservation of the reserves of . . . [the Western] . . . Hemisphere."[27] As it turned out, these hoped-for consequences were precisely what the Texas independents feared.

*

WHAT happened next makes most sense if traced separately in three parallel developments: the continuing battle with Ickes over control of policy, the actual negotiations with the British, and the bitter reaction of the Texas independents. Ickes' fierce infighting had serious consequences in the other two areas.

As will be recalled, the Petroleum Division had recommended to Hull that he secure Roosevelt's backing for State control of the talks, and an industry advisory panel to solidify that position. Accordingly, Hull, with Roosevelt's written concurrence, advised Ickes on January 3 that the American delegation would be headed by Charles B. Rayner, the newly appointed petroleum advisor in the Department of State, assisted by Paul W. Alling, chief of the Near Eastern Division, and a third member to be appointed by Ickes.[28] In a separate letter, Hull asked Ickes to discontinue further

[27] "Memorandum on the Department's Position. . . ," Sappington, December 1, 1943, folder: "Petroleum Reserves Corporation Activities, 7/3/43-1/1/44," box 1, Records of the Petroleum Division, RG59.

[28] Memorandum, Hull to Roosevelt, January 1, 1944, initialed "CH OK FDR," and letter, Hull to Ickes, January 3, 1944, unmarked black binder, box 3, Records

talks with Socal, Texas, and Gulf about possible government participation in Middle Eastern concessions, and on January 6, Rayner informed Ickes of Hull's plan to assemble an industry advisory panel himself.[29] Ickes was furious. He had already decided that he "could not use Davies on this job" (presumably because of potential conflict of interest over his Socal connections) and if he himself were to participate, the delegation would probably have to include Hull, who, as his cabinet senior, would be chairman.[30] With all of this in mind, he flatly refused to appoint a representative and appealed directly to Roosevelt to have the negotiations conducted at the cabinet rather than the technical level because of the "magnitude" of the problem.[31] He also strenuously objected to being told to discontinue his negotiations with Socal and Texas (even though they had, in fact, been in abeyance since the preceding October), and to Hull dealing directly with the oil industry.[32] Roosevelt, deeply immersed in the conduct of the war, asked the two men to resolve their differences between themselves,[33] but by the end of January it looked as though the president would have to intervene.

of the Petroleum Division, RG59; and memorandum, Hull to Roosevelt, January 5, 1944, folder: "Saudi Arabia Pipeline," box 86a, P.S.F., Roosevelt MSS; Hull, II, 1520-23.

[29] This action had been recommended by Sappington and Murray in December; memorandum, Murray to Stettinius and Hull, December 14, 1943 (drafted by "JCS"), unmarked black binder, box 3, Records of the Petroleum Division, RG59. See also, letter, Hull to Ickes, January 5, 1944, folder: "Foreign Oil Policy," box 666, item 11, RG253; and memorandum, Rayner to Hull, January 6, 1944, unmarked black binder, box 3, Petroleum Division, RG59.

[30] Diary entry for December 26, 1943, Ickes MSS.

[31] Letter, Ickes to Roosevelt, January 4, 1944, folder: "Saudi Arabia Pipeline," box 86a, P.S.F., Roosevelt MSS. See also: letter, Ickes to Hull, January 4, 1944, unmarked black binder, box 3, Records of the Petroleum Division, RG59.

[32] Letter, Ickes to Roosevelt, January 7, 1944, folder: "Saudi Arabia Pipeline," box 86a, P.S.F., Roosevelt MSS; letter, Ickes to Hull, January 7, 1944, folder: "Foreign Oil Policy," box 66, item 11, RG253; letter, Ickes to Hull, February 1, 1944, and memorandum, Raynor to Stettinius, February 2, 1944, both in unmarked black binder, box 3, Records of the Petroleum Division, RG59; and diary entry for January 9, 1944, Ickes MSS.

[33] Memorandum, Roosevelt to Hull and Ickes, January 10, 1944, 800.6363/1431, RG59.

In the midst of all this, Commodore Carter of the Army-Navy
Petroleum Board returned to Washington with a clever idea after
a tour of the Middle East. It will be recalled that it was Carter
who originally suggested a government-financed refinery for bargaining leverage with Socal and Texas. Now Carter proposed to
Ickes that the Petroleum Reserve Corporation build a pipeline
from the Persian Gulf to the Mediterranean for bargaining leverage
with the British.[34] Carter was convinced that some form of direct
government participation in Middle Eastern oil was essential to
strengthen the American position, and Socal especially was believed to be amenable to a project of this type.[35] Collier and
Rodgers were worried over the investment that would be required
to open up a postwar market for Saudi oil,[36] and having lost
government aid for the Ras Tanura refinery, they should be quite
agreeable to government financing for a pipeline to move their
oil toward the postwar European market.[37] From Ickes point of
view, however, the pipeline idea provided an excellent opportu-

[34] Diary entry for January 22, 1944, reporting a conversation with Carter on
January 17, Ickes MSS. Carter reportedly believed that "the best way to get serious
consideration from the British in the prospective negotiations . . . [was] . . . to
have some ace up our sleeve ourselves"; letter, Knox to Hull, January 11, 1944,
folder: "Petroleum Reserves Corporation Activities, 1/1-7/1/44," box 1, Records
of the Petroleum Division, RG59. Casoc (Aramco as of January, 1944) had begun
planning of such a pipeline at least as early as December 27, 1943; memorandum,
"The Position of the Department on the Petroleum Reserves Corporation" (February, 1944?), p. 15, folder: "Petroleum Reserves Corporation Activities, 1/1-7/
1/44," box 1, Records of the Petroleum Division, RG59. Carter's innovation was
to suggest that the government itself build the pipeline.

[35] Diary entry for January 22, 1944, recording a conversation with Davies on
January 18, Ickes MSS. Davies reported that the chief advocate for cooperation
with the government was Socal vice-president J. H. MacGaregill, but it is not
clear whether or not Carter discussed the specific idea of a pipeline with Socal
before he broached it to Ickes.

[36] Follis interview, San Francisco, August 24, 1977.

[37] Years later Wallace Pratt was reported to have said that "the reason Jersey
and Mobil went into Aramco [after the war] was that when the federal government
pulled out of the Saudi Arabian picture, neither California nor Texaco had enough
money to either build a pipeline or to handle other agreeements made"; letter to
the author from Professor Bennett H. Wall of Tulane University, July 13, 1977,
referring to an interview with Pratt.

nity to reassert leadership, and he moved with great vigor and dispatch to implement the plan. Although one can never be completely certain, the speed with which negotiations were completed and the fact that all of this occurred in the midst of Ickes' battle with Hull suggest that bureaucratic infighting was at least one of the motives behind the project. If the PRC were firmly committed to a Middle Eastern pipeline, then its president could scarcely be bypassed in negotiations with the British.[38] In all of this there was no advance discussion of the potential reaction of the American independents—in spite of the fact that an ugly storm had already begun to brew in Congress over the PRC. There was discussion of possible opposition from Jersey and Socony, but the discussion was based only on Carter and Ickes' opinion that those two companies had cast their lot with the British and that Jersey had an "inside track" with the State Department.[39] Ickes now saw himself allied with Socal, Texas, and Gulf against Jersey, Socony, the British, and the Department of State.

Carter proposed the plan to Ickes on Monday, January 17, and by Thursday Ickes had called in Collier and Rodgers and broached the idea to them. Collier lost no time in "agreeing that . . . [the] . . . proposal was the only way out for the California Arabian Oil Company." Rodgers was more reticent but "admitted that if they didn't get aboard this boat anything might happen."[40] Colonel Drake was called to Washington to join the negotiations for Gulf, basic agreement was reached by Saturday, and at 10:00 a.m. the following Monday a formal understanding was signed in Ickes' office.[41] Subject to later negotiation of a detailed contract, it was

[38] In all fairness, it should be noted that nowhere in his own writings does Ickes give this as his motive, but the circumstantial evidence is quite compelling. In his confidential diary, discussion of the pipeline is thoroughly interwoven with discussion of the battle with Hull. In his 1948 congressional testimony Ickes asserted that, like Carter, his motive had been to "alert the British to the idea that we really meant business in the Middle East" (U.S. Congress, *Petroleum Arrangements with Saudi Arabia*, p. 25244), but this would not be the first time that he had told a misleading half-truth.

[39] Diary entries for January 22, and 23, and February 12, 1944, Ickes MSS.

[40] Diary entry for January 23, 1944, Ickes MSS.

[41] Diary entry for January 27, 1944, Ickes MSS.

agreed that the government would construct a main trunk pipeline system from ''Saudi Arabia and Kuwait to a port at the eastern end of the Mediterranean.'' The estimated cost was between $100,000,000 and $120,000,000, to be amortized over a twenty-five-year period through fees paid by the companies for use of the system. In addition, the companies agreed to maintain a one-billion barrel reserve for potential purchase by the government at a 25 percent discount (the old option-contract idea).[42]

Carter and Ickes quickly obtained the support of the Joint Chiefs of Staff, the secretaries of war and the navy, and the director of war mobilization, and on Thursday, January 27, the document was approved by the board of the Petroleum Reserves Corporation.[43] State's representative insisted that a clause be added to the effect that ultimate ownership of the pipeline would be a matter for discussion after the war, but otherwise raised no objection. After all, a pipeline was consistent with State's basic objective of increasing Middle East production. Final approval by the president and preparation of a press release took a little longer, but the story was broken to the press on Sunday, February 6, just three weeks after the idea was first proposed.[44] As it turned out, the timing could hardly have been worse. A resolution calling for abolishment of the Petroleum Reserves Corporation had already been introduced in the Senate by Senators Edward H. Moore of Oklahoma and Owen Brewster of Maine,[45] and announcement of

[42] ''Outline of Principles of Proposed Agreement,'' January 24, 1944, bound volume, Petroleum Reserves Corporation, Records Regarding Claim Against PRC by Arabian American Oil Company, RG234.

[43] Diary entry for January 30, 1944, Ickes MSS; letter, Leahy to the secretaries of war and the navy, January 25, 1944, folder: ''CCS463.7 (4-9-43) Sect. 3,'' JCS Decimal File, 1942-45, RG218; letters, Byrnes to Roosevelt, January 17 and 25, 1944, folder: ''1943-45,'' O.F.56, Roosevelt MSS; minutes of meeting of January 27, 1944, bound volume: ''Petroleum Reserves Corporation,'' Records of Claim Against PRC by Arabian American Oil Company, RG234.

[44] Diary entries for February 6 and 12, 1944, Ickes MSS.

[45] U.S. Congress, *Congressional Record*, 78th Congress, 2d sess., vol. 90, pt. 1, pp. 494, 496-97, 1135-38. Moore was an independent oil man himself and Brewster was a leading Republican critic of the administration; Shwadran, pp. 322-23.

the pipeline plan immediately created a sustained uproar in the industry and in Congress. But we shall return to that shortly.

Emboldened by temporary success with the pipeline scheme, and taking advantage of Hull's absence from Washington during the week of February 7, Ickes next induced the president to name him chairman of the group to negotiate with the British—with the undersecretaries of state, war, and the navy, and the vice-chairman of the War Production Board as members. On learning this, Hull stoutly maintained that State should head the delegation, and pointed out that, with the president's concurrence, he had already informed the British that negotiations would be at the technical level with a delegation headed by Rayner. Roosevelt decided on a compromise on February 15, with Hull as chairman of the delegation, Ickes as vice-chairman, and the undersecretaries of war and the navy, the vice-chairman of the War Production Board, and State's petroleum advisor as additional members.[46] It appears to have been assumed by all concerned that Hull would open the first meeting and then step aside, so that de facto leadership would fall to Ickes. Ickes' fight for control had now escalated the talks to cabinet level, and when this intent was communicated to the British, it met with strong objections. An entirely unnecessary irritant had been added to the negotiations.

*

IN RETROSPECT it is difficult to know exactly what the British reaction would have been had they been quietly asked to discuss the State Department proposal without the complications created by Ickes. The working relationship between the government and the companies was much closer in London than in Washington, and when Anglo-Iranian first broached the idea of governmental talks to prevent chaos in the postwar marketing of Middle Eastern oil, the Foreign Office agreed, but preferred to have the initiative

[46] Diary entry for February 12, 1944, Ickes MSS; Hull, II, 1522-23; memorandum, Stettinius to Roosevelt, February 12, 1944, 800.6363/1495a, RG59. The others would have been Stettinius (State), Patterson (War), Forrestal (Navy), and Charles E. Wilson (War Production Board).

come from the United States.[47] In addition, when the British Chiefs of Staff examined the question in April 1944, they concluded that an agreement would have considerable strategic value because "American continental resources constitute our most secure supply in war" and expansion of Middle East production would conserve American reserves. Furthermore, American involvement in Middle Eastern oil would improve "the chances of obtaining American assistance" in defending the area, especially against possible "Russian pressure."[48] Key groups in London therefore favored the idea for their own reasons, and it can be argued that the British would have been quite amenable to discussing Middle Eastern oil without Ickes' saber-rattling. As matters turned out, however, the initial British reaction was tempered by fears that the Petroleum Reserves Corporation represented an American attempt to compete rather than cooperate in the Middle East. It took several months to put those fears to rest and get down to serious discussion of the State Department plan.

When news of Ickes' attempt to buy all or part of Casoc broke in August, 1943, the Petroleum Department of the Ministry of Fuel and Power had reacted with special concern. Its representative in the United States, Harold Wilkinson, and its chief officer in London, Sir William B. Brown, feared that British oil companies could not compete successfully with the American government, and they set out to add their influence to the attempts to block the plan. Initially, this took the form of a strong recommendation to the Foreign Office that talks be scheduled as a

[47] Foreign Office minutes, July 20, 1943, V34978, E4582/3710/65, FO/371. Sir Frederick Godber, managing director of Shell Trading and Transport, preferred to have talks between the companies first, but his view was overtaken by the course of events; Foreign Office minutes, October 1, 1943, V34210, A9286/3410/45, FO/371. For an excellent account of the Anglo-American negotiations from the British viewpoint, see Michael B. Stoff, *Oil, War, and American Security: The Search for a National Oil Policy, 1941-1947*, pp. 89-177.

[48] War Cabinet Paper W.P.(44)187, April 5, 1944, V42693, W5803/34/76, FO/371. These strategic arguments figured prominently in the cabinet's final instructions to the British delegation; War Cabinet Paper W.P.(44)179, April 5, 1944, and extract from War Cabinet Conclusion 45(44)54, April 5, 1944, both in V42693, W5619/34/76, FO/371.

means of allaying American fears and decreasing the need for such a radical move.[49] But as opposition mounted in the United States, the Petroleum Department reversed itself and argued that since domestic American opposition was well on the way to killing the idea, talks were really unnecessary.[50] A formal reply to Hull's December note was therefore delayed while bureaucrats argued over how to respond, and a decision was still pending when news broke that the PRC had decided to build a trans-Arabian pipeline, and that the proposed talks had been escalated to ministerial level. The Petroleum Department was now *sure* that the American intent was competitive and hostile. In addition, the Lord Privy Seal, Beaverbrook, the temperamental counterpart to Ickes within the British government, joined the fray in opposition to any negotiations with the belligerent Americans. "Oil," he told the War Cabinet, "is the greatest single post-war asset remaining to us," and, "[w]e should refuse to divide our last resource with the Americans."[51] The ensuing debate within the British government was masked with a protest to Washington that persons of ministerial rank could not be spared from London during preparations for the invasion of France. The matter finally required an exchange of cables between Roosevelt and Churchill to persuade the British to conduct the talks on American turf and American terms. In the process, however, it was agreed that ministerial talks would be

[49] Letters, Wilkinson to Brown, September 2, and 29, 1943, Foreign Office minute, October 1, 1943, and cable, Foreign Office to Washington, October 21, 1943, V34210, A9286/3410/45, FO/371; cable, Washington to Foreign Office, October 31, 1943, V34210, A9940/3410/45, FO/371. Churchill again decided against a British initiative for this purpose; Foreign Office minute, November 4, 1943, V34210, A10104/3410/45, FO/371.

[50] Foreign Office minute, January 17, 1944, V42686, W1361/34/76, FO/371; memorandum, "Anglo-American Oil Policy . . . prepared by the Petroleum Division for the Chairman of the Oil Control Board," January 27, 1944, V42686, W1384/34/76, FO/371; draft memorandum, J. H. LeRougetel to (British) Secretary of State, February 6, 1944, V42687, W2206/34/76, FO/371; memorandum for the prime minister, February 11, 1944, V42687, W2486/34/76, FO/371.

[51] Memorandum by the Lord Privy Seal, War Cabinet Paper W.P.(44)102, February 11, 1944, V42687, W2486/34/76, FO/371. A second memorandum by Beaverbrook is in War Cabinet Paper W.P.(44)109, February 15, 1944, V42688, W2919/34/76, FO/371.

preceded by talks at the technical level, and this was where the real negotiations finally took place.[52] It was therefore Roosevelt rather than Ickes who prodded the British into negotiations, but presidential intervention would not have been necessary in the first place if Ickes' PRC venture had not aroused deep suspicions within the British government.

Once high-level sensitivities had been dealt with or overridden, selected petroleum experts from government and industry on both sides of the Atlantic sat down to negotiate a mutually beneficial postwar arrangement. The meetings, held in Washington from April 18 through May 3, went exceptionally well. The British delegation was headed by Sir William Brown, former secretary of the Petroleum Division in the Ministry of Fuel and Power, and included Sir William Fraser, chairman of the board of Anglo-Iranian, and Sir Frederick Godber, managing director of Shell Trading and Transport (a component of Royal Dutch-Shell), along with five lesser officials from the Foreign Office, the Petroleum Division, the Treasury and the Royal Navy. The Americans were less comfortable with industry representatives serving as members of the official delegation and they used a two-tier approach. The delegation itself was headed by Charles Rayner, petroleum adviser to the Department of State, and included Ralph Davies, deputy petroleum administrator for war, Paul Alling, now deputy director of the Office of Near Eastern and African Affairs in the Department of State, Commodore Andrew Carter, executive officer of

---

[52] Memorandum, British embassy to Department of State, February 18, 1944, and memorandum of conversation, Stettinius, Halifax et al., February 18, 1944, *FR(1944)*, III, 97-99. The British fully understood that escalation of the talks to ministerial level represented a victory for Ickes over the Department of State, and this was one of the reasons they balked. Ickes wanted to fight it out, and State preferred cooperation; cable, Halifax to Foreign Office, February 15, 1944, V42687, W2433/34/76, FO/371; and memorandum, Eden to War Cabinet, War Cabinet Paper W.P.(44)119, February 19, 1944, V42688, W2920/34/76, FO/371; cables, Churchill to Roosevelt, February 20, 1944, Roosevelt to Churchill, February 22, 1944, Churchill to Roosevelt, February 24, 1944, Churchill to Roosevelt, March 4, 1944, Winant to Roosevelt, March 6, 1944, and Stettinius to Winant, March 7, 1944, *FR(1944)*, III, 100-105; and cable, Roosevelt to Churchill, March 3, 1944, V42690, W3697/34/76, FO/371.

the Army-Navy Petroleum Board, and six others from the War Department, the State Department, and the Petroleum Administration for War. For advice during the talks, the Americans assembled a group of ten industry executives, and three of those sat in the actual negotiations during eight of the ten meetings. The three were John Brown of Socony-Vacuum, "Star" Rodgers of The Texas Company, and Alvin Jacobsen, president of Amerada Petroleum. The others, who were consulted at the beginning and end of the talks, were H. C. Collier of Standard Oil of California, Colonel J. Frank Drake of Gulf Oil, Orville Harden, vice-president of Standard Oil (New Jersey), W. Alton Jones, president of Cities Service, William R. Boyd, Jr., president of the American Petroleum Institute, Ralph T. Zook, president of the Independent Petroleum Association of America, and George A. Hill, Jr., president of the Houston Oil Company and an active member of the Independent Petroleum Association.[53] Taken together, this was an impressive assemblage of oil men from both sides of the Atlantic. Conspicuously absent were any representatives of the Mid-

[53] Minutes of Joint Sessions [ten], Anglo-American Exploratory Talks on Petroleum, April 18 through May 3, 1944, V42695, W7397/34/76, and V42696, W8275/34/76, FO/371; Frey and Ide, p. 280; and *FR(1944)*, III, 108, n.17. The rest of the British delegation included J. H. LeRougetel, head of the General Department of the Foreign Office, Frederick C. Starling and Victor Butler of the Petroleum Division, Frederick Harmer, an assistant secretary of the treasury, and Commodore A. W. Clarke of the British Joint Staff Commission in Washington. The rest of the American delegation included Brigadier General H. L. Peckham of the War Department, Leroy D. Stinebower of the Office of Economic Affairs and James C. Sappington and John A. Loftus of the Office of the Petroleum Advisor in the Department of State, George Walden, special assistant to the deputy petroleum administrator for war, and C. S. Snodgrass, director of the Foreign Refining Division of the Petroleum Administration for War. See also Minutes of meetings of United States Technical Group, April 20 through May 3, 1944, red binder: "Petroleum Discussions with the British (except Joint Minutes)," box 3, Records of the Petroleum Division, RG59; and Frey and Ide, p. 280. The advisory panel grew out of Hull's February meeting with corporate executives, originally designed as a counter to Ickes; memoranda, Rayner to Hull, January 17, and February 1, 1944, unmarked black binder, box 3, Records of the Petroleum Division, RG59; minutes of meeting of the President's Committee on Oil, April 10, 1944, 800.6363/1614, RG59.

dle East producing countries whose oil was about to be divided up.[54]

To the surprise of the British, Rayner opened the talks by offering a complete "Memorandum of Understanding" as the basis for discussion, and the two delegations suddenly found themselves in essential agreement.[55] The document was a more sophisticated version of the government cartel idea proposed by Thornburg and Feis. Its objective was still maximization of Middle Eastern production to conserve Western Hemisphere reserves, and its key element was still a joint Anglo-American petroleum commission. But this version opened with a statement of general principles, omitted specific reference to the Middle East, theoretically defined American domestic production as outside its scope (by limiting it to "petroleum in international trade"), and called for an expanded multinational commission at some later date.[56] All of these changes were designed to allay the fears of one group or another, but the two delegations clearly understood that management of postwar Middle East production was the central issue. They were in near consensus that rational cooperation

[54] The question of equity to the various producing countries was a consideration during the talks, but from the perspective of avoiding nationalization of concessions. In all fairness, it should be noted that there was informal discussion of including producing countries in talks at a later date; cable, Halifax to Foreign Office, April 16, 1944, V42693, W5898/34/76, FO/391.

[55] Except as otherwise noted, this discussion of the technical-level negotiations is based on minutes of Joint Sessions, April 18 through May 3, 1944, V42695, W7397/34/76, and V42696, W8275/34/76, FO/371. When instructions to the British technical delegation were discussed by the War Cabinet, it was assumed that the Americans would be poorly prepared and that the talks would be only a general exchange of ideas; War Cabinet Paper W.P.(44)179, April 5, 1944, and extract from War Cabinet Conclusion 45(44)54, April 5, 1944, V42693, W5619/34/76, FO/371.

[56] There is an excellent analysis of the American objectives in the memorandum: "The Petroleum Division," October, 1944, pp. 15-36, box 48, Harley Notter files, RG59. In late February State Department planners had agreed upon a "broad policy of conservation of Western Hemisphere reserves" through "full development of Middle Eastern . . . production," but they had concluded that the logical corollary, "a program of Government-sponsored net imports of petroleum from the Middle East to the United States would, because of the political and economic difficulties involved, be unwise"; "The Petroleum Division," p. 21.

would be better than unbridled competition, and offstage protests to the contrary by Hill and Zook of the American independents went almost unnoticed.[57]

Much of the discussion centered on the proposed Joint Petroleum Commission. The Americans originally wanted it to have limited executive powers, but after several days of discussion they agreed to restrict it to an advisory role on the strength of Sir William's argument that the real power "of the Commission would derive from . . . the calibre of the people making it up, . . . and . . . the confidence it would be able to inspire throughout the petroleum industry."[58] It was agreed that each nation would appoint five members of the commission, and the British announced that at least three of theirs would be senior industry executives who would retain their corporate affiliations. The State Department believed that appointing members from industry rather than government would be unwise, but left that issue for resolution at a later date.[59] As to specific issues that the commis-

[57] Prior to the initialing of the draft Memorandum of Understanding, a copy was given to all of the industry advisors for comment. In a letter dated April 29 Hill raised serious objections to wording that might give the American government legal justification for increasing its control of domestic producers. His points were discussed but not acted upon by the American delegation; minutes, meeting of United States Technical Group, May 3, 1944, red binder: "Petroleum Discussions with the British (except Joint Minutes)," box 3, Records of the Petroleum Division, RG59. Zook was more diplomatic, but he wrote Rayner on May 5 with a long list of recommended word changes to clearly protect the American independents; letter, Zook to Rayner, May 5, 1944, 800.6363/1666, RG59. These criticisms were a foretaste of the bitter attack that followed later.

[58] Minutes of Joint Session no. V, April 22, 1944, V42696, W8275/34/76, FO/ 371.

[59] "Report on Petroleum by the Expert Technical Group to the President's Committee on Petroleum," May 10, 1944, red binder: "Petroleum Discussions with the British (except Joint Minutes)," box 3, and memorandum, Rayner to Hull, May 27, 1944, folder: "Memos re: Petroleum, 1943-44," box 1, Records of the Petroleum Division, RG59. Rayner had wanted at least two American members from industry, to facilitate "voluntary compliance with the Commission's recommendations." But Acheson had objected on the grounds that such a commission might "fail to be guided exclusively by wise national policy" and become instead "an international cartel." He also believed that "to have international petroleum policies laid down by a Commission" two of whose members

sion might want to deal with, there was considerable off-the-
record discussion of the Red Line Agreement, the Anglo-Iranian/
Gulf Kuwaiti contract, and the PRC pipeline project. The implication was that these thorns in everyone's flesh could best be dealt
with by quiet negotiations within the confines of the new commission.

On two issues, however, the Americans would not accede to
the British. Before leaving London, the British delegation had
been directed by the War Cabinet to seek an agreement for joint
action in cases such as the prewar Mexican nationalization of
British and American petroleum concessions.[60] When this was
discussed, the American delegation agreed to "respect" but not
to "support" one another's concession rights. An agreement to
"support" was seen as too large a commitment to make in advance.[61] Of greater consequence later was disagreement over a
clause to protect Britain's right to restrict its postwar imports to
oil paid for in sterling rather than dollars, if required by a serious
shortage of foreign exchange. This was a clause proposed by the
British treasury but opposed by the Department of State as contrary
to its longstanding opposition to preferential commercial agreements. The Americans argued that this was a "relevant economic
factor" that could be taken into consideration by the new commission in making its recommendations. The British delegation
finally acquiesced, and the special protective clause was not

represented "companies having a large financial interest in the outcome" did not
"seem to be sound government procedure." Acheson's view prevailed in intradepartmental discussions after the talks, but the question became moot when the
agreement failed to be ratified; memorandum, "Composition of United States
Membership in the Proposed Joint Petroleum Commission," May 25, 1944, unmarked black binder, box 3, Records of the Petroleum Division, RG59. The
British delegation was not aware of this offstage debate and reported to London
that the Americans would probably want three government and two industry
members; minutes of War Cabinet Ministerial Oil Committee meeting, May 8,
1944, V42696, W7877/34/76, FO/371.

[60] War Cabinet Paper W.P.(44)179, April 5, 1944, and extract from War Cabinet
Conclusion 45(44)54, April 5, 1944, V42693, W5619/34/76, FO/371.

[61] Memorandum, Rayner to Hull, April 26, 1944, folder: "Memos re: Petroleum, 1943-1944," box 1, Records of the Petroleum Division, RG59.

added. The Memorandum of Understanding was initialed by the heads of the two delegations on May 3 as a recommendation to their respective governments to be discussed shortly at the ministerial level.[62] The direct participants in the negotiations departed well pleased.

Rodgers was especially fulsome in his congratulations to Rayner for a job well done,[63] and John Brown added a quiet postscript indicative of his longer-term thinking. On May 15 he wrote Rayner suggesting that any congressional enabling act recommended to carry out the terms of the proposed agreement include a clause to the effect that "whenever the Government of the United States approves any recommendation or suggestion by the Commission, acts taken pursuant thereto shall be deemed in the public interest and no prosecution or civil action shall be commenced with reference thereto under the Anti-trust laws of the United States or the Federal Trade Commission Act."[64] Brown's letter probably represented the high point of support for the agreement by the international segment of the industry, but it was a contingent support. If the government really wanted to establish a binational commission to manage Middle Eastern production, it would receive the full cooperation of the majors—provided they were protected from other arms of government that had been longstanding foes of such "collusion." The question of antitrust protection proved to be a major sticking point in the debates that were to follow.

The American delegation for talks at the ministerial level had been designated and waiting in the wings since February,[65] but the British had deferred selecting delegates until after the technical-level talks had been completed. It was just as well, because

[62] Letter, Hull to Roosevelt, May 4, 1944, with "Draft Memorandum of Understanding with the United Kingdom on Petroleum," April 29, 1944, *FR(1944)*, III, 111-15.

[63] Letter, Rodgers to Rayner, May 5, 1944, 800.6363/1646, RG59.

[64] Copy of letter, Brown to Rayner, May 15, 1944, unmarked black binder, box 3, Records of the Petroleum Division, RG59.

[65] Letter, Acting Secretary of State to Halifax, February 17, 1944, *FR(1944)*, III, 95-96.

Sir William's report created a furor within the British cabinet. The cabinet committee charged with oversight of the talks was well satisfied,[66] but Beaverbrook raised strong objections. He appears to have been generally unhappy over descending British power vis-à-vis the Americans,[67] but he selected the question of "sterling oil" as his principal point of attack. The failure to specifically guarantee Britain's right to regulate its own imports was "quite unacceptable," and he vehemently opposed the agreement.[68] Foreign Minister Anthony Eden countered that failure to complete the process at this point could cause the Americans to react "irresponsibly" at a time when it was in Britain's interest to avoid "unrestricted competition." He raised the specter of the PRC pipeline to drive home the point.[69] The ensuing argument lasted almost a month and was finally resolved when Churchill appointed Beaverbrook himself to lead the ministerial delegation, with three members of the ministerial oil committee as members.[70] In effect, Beaverbrook was charged with seeing whether he could

[66] Minutes of War Cabinet Ministerial Oil Committee, May 8, 1944, V42696, W7877/34/76, M.O.C.(44)8, May 15, 1944, V42696, W7889/34/76, and minutes of War Cabinet Ministerial Oil Committee, May 16, 1944, V42696, W7970/34/76, FO/371. The committee consisted of the minister of war transport (Lord Leathers), the minister of state (Richard Law), the chairman of the Oil Control Board (Geoffrey Lloyd), the financial secretary to the treasury (Ralph Assheton), the Fourth Sea Lord (Vice Admiral A.F.E. Palliser), and the permanent undersecretary of the India Office (David T. Monteath). Sir William went so far as to recommend that the final negotiations be completed as soon as possible by the British ambassador already in Washington; M.O.C.(44)11, May 23, 1944, V42696, W7970/34/76, FO/371.

[67] Letter, Ronald I. Campbell (Washington) to Sir Orme Sargent (London), August 22, 1944, reporting a long conversation with Beaverbrook on July 28, V42702, W13390/34/76, FO/371.

[68] Memorandum, LeRougetel to Eden, May 31, 1944, V42697, W8798/34/76, FO/371.

[69] Letter, Eden to Beaverbrook, June 12, 1944, V42697, W8879/34/76, FO/371.

[70] War Cabinet Paper W.P.(44)313, June 14, 1944, V42697, W10105/34/76; War Cabinet Papers W.P.(44)324, June 14, 1944, and W.P.(44)328, June 15, 1944, V42697, W9880/34/76; cable, Churchill to Roosevelt, June 21, 1944, V42698, W10568/34/76, FO/371; minutes of Plenary Session no. 1, Anglo-American Conversations on Petroleum, July 25, 1944, *FR(1944)*, III, 119.

negotiate something better, but he was escorted by supporters of the agreement to make sure that no serious damage was done. It was a deft move, but the American government was unaware of this behind-the-scene drama and expressed considerable concern over the delay.[71]

When the two delegations finally met in Washington on July 25, the roster was impressive. The British were represented by the Lord Privy Seal, Beaverbrook, Minister of State Richard Law, Financial Secretary to the Treasury Ralph Assheton, Parliamentary Secretary to the Ministry of Fuel and Power Geoffrey Lloyd, and Minister Resident in Washington Ben Smith, with Sir William Brown as chief advisor. Hull headed the American delegation, but after the first meeting illness forced him to yield that role to Ickes, who had been appointed vice-chairman. The others were Secretary of the Navy Forrestal, Undersecretary of War Patterson, Foreign Economic Administrator Leo Crowley, Deputy Petroleum Administrator Ralph Davies, and Petroleum Advisor Charles Rayner, with Harry C. Hawkins of State's Office of Economic Affairs as an advisor.[72]

With all of the preparatory work that had been done, the final negotiations might have been expected to be almost pro forma, but, in fact, the debate was so intense that the talks almost ended in deadlock. Beaverbrook argued that their anticipated postwar foreign exchange shortages made absolutely necessary an additional clause protecting "the right of each country to draw its consumption requirements, to the extent that may be considered necessary, from the production in its territories or in which rights are held by its nationals."[73] The Americans protested that such

---

[71] Cables, Stettinius to Winant, June 6, 1944, and Roosevelt to Churchill, June 7, 1944; memorandum, Rayner to Hull, June 16, 1944; cables, Hull to Winant, June 24, 1944, and Winant to Hull, July 3, 1944; *FR(1944)*, III, 116-19. Shwadran (p. 329) did not have access to Foreign Office documents and attributed the delay to British unhappiness over the pipeline project. The documents, however, make clear that the sole reason for the delay was Beaverbrook's objection.

[72] Minutes of Plenary Session no. I, July 25, 1944, *FR(1944)*, III, 119-21. Sappington served as executive secretary of the American delegation.

[73] Memorandum, "The Petroleum Division," October, 1944, p. 35, box 48, Harley Notter files, RG59.

a clause would permit the exclusion of "dollar oil" from Britain, and even if they were to agree to it, a document with such a clause would meet with intense political opposition within the United States. The argument dragged on for almost two weeks, amid extensive cabled consultation with the British cabinet, but the Americans held firm. Finally, a compromise was struck. The protective clause was not added, but it was agreed to record both positions in the official minutes, and the British reserved the right to terminate the understanding on three months notice if an actual problem of this type could not be satisfactorily resolved in the future by the new Petroleum Commission.[74] In addition, the two delegations agreed to reduce the number of representatives on the commission to four each, from government only, with supporting advisory groups from industry on both sides of the Atlantic. The problem of "sterling oil" sidestepped in 1944 was not finally resolved until Standard Oil (New Jersey) negotiated a special agreement directly with the British government in 1950.[75] But the 1944 compromise avoided a breakdown in the Washington negotiations, and an "Anglo-American Petroleum Agreement" was signed by the two delegations on August 8. The complete text is given in Appendix B. With only a few minor changes it was the same document agreed upon in the technical-level negotiations three months earlier.

    After some last-minute debate within the State Department over

[74] Ibid.; minutes of plenary sessions and meetings of Joint Subcommittee, Anglo-American Conversations on Petroleum, July 26-August 3, 1944, V42700, W12096/34/76 and V42701, W12843/34/76; cable, Campbell to Foreign Office, August 2, 1944, V42699, W11844/34/76; extract from War Cabinet Conclusion 100(44), August 3, 1944, V42699, W12003/34/76; cable, Campbell to Foreign Office, August 4, 1944, V42699, W11910/34/76; cable, Campbell to Foreign Office, August 10, 1944, V42700, W12113/34/76; letter, Halifax to Eden, August 24, 1944; and letter, Campbell to LeRougetel, August 30, 1944, V42701, W13044/34/76, FO/371.

[75] Britain's exchange position became critical in 1949 and Jersey's sales in Britain were hurt by various regulations imposed by the British government. Among the interests adversely affected were those of Aramco and Saudi Arabia, because Saudi production was classified by Britain as "dollar oil." For discussion of how this problem was resolved, see Chapter VI.

whether the document should be handled as an executive agreement or a treaty, the department yielded to pressure from the Senate Foreign Relations Committee and through the president the agreement was transmitted on August 24 as a treaty to the Senate for its "advice and consent."[76] Its phraseology was now somewhat hazy, but internal State Department documents make clear that the long-run objective was "to effect a change in the geographic pattern of production" in order to conserve Western Hemisphere reserves. The immediate objective was "to improve . . . relations with the British . . . so that petroleum resources financially controlled by United States nationals could be produced on a larger scale without restrictions and interferences." And the permanent mechanism for accomplishing all of this was the "joint Anglo-American Petroleum Commission," which would "recommend production and exportation rates for the various concession areas in the Middle East . . . [to prevent] . . . the disorganization of world markets which might result from uncontrolled competitive exploitation."[77] As an abstract proposition, this approach had some merit. It was basically supported by the British government, the British oil companies, and the American majors. Had it come into operation, it might have provided a mechanism for government oversight and protection of long-term consumer interests. And if it had been expanded to include representation for the producing countries (as was envisioned), it might have

[76] Memorandum, "The Petroleum Division," October, 1944, pp. 38-39, box 48, Harley Notter files; memorandum of conversation, Assistant Secretary of State Breckinridge Long, Rayner, and members of the Senate Foreign Relations Committee, plus Senators Francis Maloney and E. H. Moore (of the Special Committee Investigating Petroleum), August 16, 1944, 800.6363/8-1644; memorandum, Rayner to Hull, August 17, 1944, 800.6363/8-1744; memorandum of telephone conversation, Long and Connally, August 24, 1944, 800.6363/8-2444, RG59; letter, Hull to Roosevelt, August 24, 1944, *FR(1944)*, III, 124-25; and *Congressional Record*, 78th Congress, 2d sess., August 25, 1944, vol. 90, pt. 6, pp. 7303-5. Acheson strongly favored holding to the original plan to treat it as an executive agreement, but Long's talks with several key Senators convinced him that that would be politically unwise. Hull decided in favor of Long.

[77] Suggested Memorandum . . . for the Secretary, by John A. Loftus (Petroleum Division), November 9, 1944, folder: "Memos re: Petroleum, 1943-1944," box 1, Records of the Petroleum Division, RG59.

made unnecessary the later formation of The Organization of Petroleum Exporting Countries (OPEC) to protect those interests as well. But the approach ran counter to the vested interests of the American independents, the antitrust philosophy of the Department of Justice, the laissez-faire ideology of a remnant of New Deal opponents, and State's long-standing practice of not supporting one domestic interest group over another. Whether the now somewhat ambiguous agreement could have been safely navigated through these shoals in quiet waters will remain a matter for speculation. In fact, the idea ran aground in the domestic storm created by the Petroleum Reserves Corporation.

*

INDUSTRY concern over the secretive actions of the PRC had begun in October, 1943, when word leaked out that an attempt was under way to purchase a controlling interest in the Saudi concession. As already noted, the Foreign Operations Committee of the Petroleum Administration for War immediately took a vehement stand against direct government participation in the oil business, called for diplomatic support of private enterprise abroad, and proposed an international agreement similar to the Interstate Oil Compact of 1935. This statement was taken up by the December meeting of the Petroleum Industry War Council (PIWC), representing the entire industry, and an ad hoc Committee on Foreign Oil Policy was appointed to study it.[78] Since the original statement had been prepared by representatives of companies engaged in foreign operations, the PIWC agreed that the ad hoc committee should consist solely of those engaged in domestic operations, and George Hill and Ralph Zook of the Independent Petroleum Association of America (IPAA) were among those appointed. Two days later the IPAA itself adopted a strongly worded resolution calling on the government to give diplomatic support to private enterprise overseas, but not to "directly or indirectly engage in foreign oil ownership, exploration, development, or op-

[78] Minutes of meeting, Petroleum Industry War Council, December 8 and 9, 1943, folder: "Petroleum Industry War Council," box 687, item 15, RG253.

eration'' in any way, shape, or form.[79] When the PIWC ad hoc committee reported in January, the PIWC voted to endorse the original statement of the Foreign Operations Committee, except for the section on the interstate oil compact, on which it reserved judgment. In addition, it went on record as solidly supporting the statement of the Independent Petroleum Association. Ickes was taken aback by all of this and requested a further study of the entire matter by the PIWC. To do this, the PIWC established a new National Oil Policy Committee with John Brown of Socony-Vacuum as chairman. Its membership included George Hill, Ralph Zook, Colonel Drake of Gulf Oil, and Eugene Holman, vice-president of Standard Oil (New Jersey), among others.[80] On February 2, this committee proposed, and the PIWC adopted, a resolution calling for the immediate dissolution of ''the Petroleum Reserves Corporation to the end that a considered and sound national oil policy may be formulated.''[81] By the first of February 1944, suspicion and opposition had reached a high level of intensity in all segments of the industry except for Socal, The Texas Company, and Gulf, who maintained as low a profile as possible.

Meanwhile, the issue had boiled over into the United States Senate. On January 21, Senator E. H. Moore, himself an independent oil man from Oklahoma, read into the *Congressional Record* the full text of the resolutions adopted by the PIWC and IPAA, and in conjunction with Senator Owen Brewster, an antiadministration Republican from Maine, he introduced a resolution calling for liquidation of the Petroleum Reserves Corporation.[82] Brewster was one of a group of five senators who had made a round-the-world inspection tour of wartime petroleum operations under the auspices of the Truman Committee in the summer of 1943, and who had returned seriously concerned over

[79] Ibid.

[80] Ibid.; most of the PIWC resolutions and documents cited in the following account appear also in U.S. Congress, *Documentary History of the Petroleum Reserves Corporation*, pp. 59-119.

[81] John W. Frey and H. Chandler Ide, *A History of the Petroleum Administration for War, 1941-1945*, p. 277; diary entry for February 6, 1944, Ickes MSS.

[82] U.S. Congress, *Congressional Record*, 78th Congress, 2d sess., January 21, 1944, vol. 90, pt. 1, pp. 494-97; diary entry for February 6, 1944, Ickes MSS.

the depletion of American reserves and convinced that Congress
should play an active role in the development of a national oil
policy.[83] On February 3, Moore delivered a blistering attack
against the Petroleum Reserves Corporation on the floor of the
Senate. Among other things, he charged that it was part of a long-
term plot by "socialistic and communistic-minded New Dealers"
to take over control of the oil industry. The alleged decline in
proven American reserves was a "man-made shortage" brought
about by price controls that had driven independent wildcat op-
erators from the field. This "repression of prices" would "eventually bring about a . . . scarcity of crude oil" and "reduce the
industry to a few large units, which . . . [could] . . . be easily
taken over or governmentally controlled, and . . . [would] . . .
at the same time arouse public opinion in support of engaging this
nation in the oil business abroad." He singled out Ickes as among
those who supported "such international socializing of petroleum," and called upon "the people of our Nation . . . [to]
. . . demand a cessation of further encroachment of Government
into business."[84]

Such was the climate of opinion in the industry and the Senate
when Ickes made the surprise announcement on February 6 that
the Petroleum Reserves Corporation had entered into an agreement
with Socal, The Texas Company, and Gulf to build a pipeline at
government expense across the Arabian peninsula. The effect
could not have been more provocative if Ickes had specifically

[83] U.S. Congress, Senate, Special Committee Investigating the National Defense
Program, *Report of Subcommittee Concerning Investigations Overseas, Section
1, Petroleum Matters*, pp. 1-16. The five were Brewster and James M. Mead of
New York (appointed by Senator Harry S. Truman, chairman of the Special
Committee Investigating the National Defense Program), A. B. Chandler of Ken-
tucky and Henry Cabot Lodge, Jr. of Massachusetts from the Military Affairs
Committee and Richard B. Russell of Georgia from the appropriations Committee.
Immediately after their return all five had issued public statements that contributed
to rising concern throughout the country over depletion of American reserves.
Their formal report was delivered to the Senate on February 16, 1944 and added
impetus to the drive for an investigation of the petroleum reserves question.

[84] U.S. Congress, *Congressional Record*, 78th Congress, 2d sess., February
3, 1944, vol. 90, pt. 1, pp. 1135-38.

devised a plan to antagonize both Congress and the industry. Three days after the announcement Brewster and Moore introduced a resolution calling for a special committee representing the Committees on Foreign Relations, Commerce, and Interstate Commerce to investigate the pipeline proposal and "all similar proposals from top to bottom."[85] Their resolution evolved into the Special Committee to Investigate Petroleum Resources, initially chaired by Senator Francis Maloney of Connecticut, which scheduled its first public hearings for the first part of May.[86] Representative Roy O. Woodruff of Michigan attacked the pipeline in the House as a "step forward toward Government-owned, Government-controlled, and Government-regimented industry, and toward a national economy based on Marxist economic principles."[87]

Politically most potent, however, was a lengthy analysis of the pipeline scheme drafted by George Hill, forwarded by Ralph Zook of the IPAA to John Brown of the PIWC's National Oil Policy Committee, and endorsed by the PIWC at its March meeting, along with a second resolution calling for dissolution of the PRC.[88] Among other things, Hill charged that "citizens of the United States now find themselves confronted with a corporation, organized without any specific authorization of Congress, conferring

[85] Ibid., February 9, 1944, vol. 90, pt. 2, pp. 1466-71.

[86] Ibid., April 21, 1944, vol. 90, pt. 3, pp. 3615-16. Maloney was subsequently replaced as chairman by Senator Joseph C. O'Mahoney of Wyoming, and the committee held extensive hearings in 1945 and 1946. It was a distinguished group, including, among others, Brewster, Moore, Tom Connally of Texas, Arthur H. Vandenberg of Michigan, and Robert M. LaFollette, Jr. of Wisconsin; U.S. Congress, Senate, Special Committee Investigating Petroleum Resources, *Intermediate Report*, frontspiece.

[87] Remarks entered into the Record for March 24, 1944; U.S. Congress, *Congressional Record, Appendix*, 78th Congress, 2d sess., vol. 90, pt. 8, p. A1469.

[88] Minutes of meeting, Petroleum Industry War Council, March 1 and 2, 1944, folder: "Petroleum Industry War Council," box 687, item 15, RG253; and *United States Foreign Oil Policy and Petroleum Reserves Corporation: An Analysis of the Effect of the Proposed Saudi Arabian Pipe Line* (Washington, D.C.: Petroleum Industry War Council, 1944), copy in 800.6363/1732, RG59; diary entry for March 5, 1944, Ickes MSS.

upon itself . . . powers overlapping and duplicating functions of several departments of government . . . , expending funds not appropriated for such purpose by . . . Congress . . . , pursuing an objective contrary to the American way of life and of private enterprise by engagement in the oil pipeline business . . . , [and] . . . making the most dangerous commitments in the field of foreign policy.''[89] To make sure that the message was heard, the PIWC distributed ninety thousand copies of the Hill analysis,[90] and scores of newspaper editorialists picked up the same themes.[91]

Herbert Feis, however, put his finger on the heart of the matter. In an analysis published in March 1944, he pointed out that the proposed pipeline would ''affect the competitive position of each and every oil-producing interest . . . [and] . . . the other American oil companies . . . regard themselves as exposed to what they feel is favored competition.''[92] The outpouring of rhetoric, in fact, originated with domestic independents and nonparticipating majors who appealed to American laissez-faire ideology and created a storm of opposition in defense of their interests.[93] Direct government assistance to Socal, Texas, and Gulf in moving low-cost Saudi and Kuwaiti oil closer to European, and potentially American, markets was simply unacceptable to the rest of the industry. Ickes' stock purchase plan and pipeline proposal had thus created an organized opposition centered on George Hill and Ralph Zook's aroused Independent Petroleum Association of America, John Brown's National Oil Policy Committee of the Petroleum Industry

[89] Ickes MSS, p. 7.   [90] Frey and Ide, p. 279.

[91] The British embassy provided London with an extensive analysis of American press comment; see, for example, letter, Halifax to Eden, April 11, 1944, V42694, W6463/34/76, FO/371.

[92] Herbert Feis, *Petroleum and American Foreign Policy*, p. 45; Feis' criticism of the pipeline proposal was considered quite significant; *New York Times*, April 7, 1944.

[93] Rodgers and Drake told Ickes that representatives of Jersey and Socony had suggested lines of attack to the *New York Times* and *Herald Tribune* (diary entries for March 5, 11, and 20, 1944, Ickes MSS). Whether or not such direct stimulation of press criticism actually occurred, the indirect stimulation through the PIWC was more than adequate to account for the storm of adverse comment that spread across the country.

War Council, and Senator Francis Maloney's Special Committee to Investigate Petroleum Reserves. This was a formidable alignment. It is interesting to note that both the independents and the nonparticipating majors were opposed to the pipeline scheme, but in general the majors let the independents lead the public outcry. The reason for this is not explicit in the accessible record, but it can be speculated that because the majors had been under heavy attack in Congress in the early 1940s for monopolistic practices, they preferred to keep as low a profile as possible on controversial issues such as this.

The pipeline plan now fell on hard times. Senator Maloney attempted to get Ickes to agree not to sign a final contract with Socal, Texas, and Gulf without at least thirty days' prior notice in order to give his committee time to study the question and act if necessary. He also offered to hold off on public hearings until after the Anglo-American talks were completed in order to avoid anti-British publicity if Ickes would cooperate. Ickes would agree to only ten days' notice, and Maloney finally went to Roosevelt, who gave the committee the necessary thirty days' assurance.[94] In the midst of this controversy, Ickes turned to the Joint Chiefs of Staff for reaffirmation of their earlier support of the pipeline as "a matter of immediate military necessity."[95] Since it was now anticipated that the war in Europe would be over before the pipeline was completed, the Joint Chiefs declined such endorsement.[96]

[94] Minutes of meeting of directors, Petroleum Reserves Corporation, May 12, 1944, bound volume: "Petroleum Reserves Corporation," Records Regarding Claim Against PRC by Arabian American Oil Company, RG234; letters, Byrnes to Roosevelt, May 22, 1944, Roosevelt to Ickes, May 24, 1944, Ickes to Roosevelt, May 29, 1944, Roosevelt to Ickes, May 30, 1944, Roosevelt to Watson, May 31, 1944, Maloney to Byrnes, June 7, 1944, Byrnes to Roosevelt, June 9, 1944, Roosevelt to Maloney, June 12, 1944, and Maloney to Roosevelt, June 14, 1944, folder: "Saudi Arabia Pipeline," box 86a., P.S.F., Roosevelt MSS.

[95] Letter, Ickes to Forrestal, June 8, 1944, folder: "CCS463.7 (4-9-43) Sect. 3," JCS Decimal File, 1942-45, RG218; see also minutes of meeting of directors, Petroleum Reserves Corporation, May 12, 1944, bound volume: "Petroleum Reserves Corporation," Records Regarding Claim Against PRC by Arabian American Oil Company, RG234.

[96] Joint Chiefs of Staff 281/17, "Pipeline from the Arabian Oil Fields to the Mediterranean," June 20, 1944; excerpt from minutes of Joint Chiefs of Staff,

To further complicate matters, the signing of a definitive pipeline contract with Socal, Texas, and Gulf continued to be delayed because Drake, Rodgers, and Collier could not agree on the proportion of the line that would be allocated to Gulf.[97] That would not have been an insurmountable problem, but by this time the Anglo-American technical negotiators had agreed that the pipeline would be one of the items referred to the new binational Petroleum Commission. In the face of all these setbacks, Ickes now began to emphasize in private conversations that the *real* purpose of the pipeline proposal had been to get the British to talk, and that from this viewpoint it had been eminently successful.[98] What is more likely is that Ickes' prime objective had been to win center stage for himself in the direction of foreign oil policy, and, if so, the manuever had worked. When the ministerial conference assembled in August and Hull stepped aside due to illness, Ickes became acting head of the American delegation and completed the final negotiations. He never appears to have understood that his tactics delayed rather than accelerated an agreement with the British, and that they created such suspicion of the government's objectives that the agreement itself was now in jeopardy. Be that as it may, it is clear that by the end of June Ickes had allowed the pipeline scheme to lapse. No further meetings of the PRC board were held; Socal and Texas decided to start work on the pipeline themselves; and the PRC was liquidated along with the remaining assets of the Reconstruction Finance Corporation after the end of the war. Its legacy was the Maloney Committee and a suspicious and aroused petroleum industry.

*

June 28, 1944; and memorandum, Chief of Staff, U.S. Army to Secretaries of War and the Navy, June 28, 1944; folder: "CCS463.7 (4-9-43) Sect. 3," JCS Decimal File, 1942-43, RG218.

[97] Diary entries for April 9 and May 14, 1944, Ickes MSS; minutes of meeting of directors, Petroleum Reserves Corporation, May 12, 1944, bound volume: "Petroleum Reserves Corporation," Records Regarding Claim Against PRC by Arabian American Oil Company, RG234. Drake wanted one-third for Gulf and Rodgers and Collier would only agree to one-fourth.

[98] Ibid.; diary entry for May 7, 1944, Ickes MSS.

THE STATE DEPARTMENT had originally planned to handle the Anglo-American Agreement as an executive agreement not requiring ratification, but it had agreed to fully inform the Senate Foreign Relations Committee of all that had transpired. When Rayner and Assistant Secretary of State Breckinridge Long met with the committee for that purpose on August 16, they were surprised to find that the chairman, Senator Connally, had invited Senators Maloney and Moore of the Special Petroleum Committee to join them.[99] In the discussion that followed it developed that all but one of the twelve senators present was critical of the agreement, and that a majority thought that it shoud be dealt with as a treaty. In retrospect, Long and Rayner were convinced that the cold reception stemmed at least in part from a highly critical analysis distributed to the senators in advance by legal counsel for the Special Petroleum Committee. After considerable debate within the department, Hull concurred with Long's recommendation that the agreement be handled as a treaty, and it was transmitted to the Senate by presidential letter on August 24.[100] This move may have been a necessary one in the atmosphere that then prevailed, but the effect was disastrous.

As already noted, State's original objective had been a bilateral commission to manage Middle Eastern production and ensure market outlets for Saudi oil.[101] In this, it had had the support of the majors, provided they would be protected against antitrust

[99] Memorandum of conversation, Long, Rayner, and Connally et al., August 16, 1944, 800.6363/8-1644, and memorandum, Rayner to Hull, August 17, 1944, 800.6363/8-1744, RG59.

[100] Memorandum of telephone conversation, Long and Connally, August 24, 1944, 800.6363/8-2444, RG59; letter, Hull to Roosevelt, August 24, 1944, *FR(1944)*, III, 124-25; and letter, Roosevelt to Senate, August 24, 1944, *Congressional Record*, 78th Congress, 2d sess., August 25, 1944, vol. 90, pt. 6, pp. 7304-5.

[101] The following commentary is based, in part, on a highly insightful analysis prepared by John A. Loftus of the Petroleum Division in the fall of 1944; see memoranda, Sappington to Rayner, October 30, 1944, and "Background Information on the Anglo-American Oil Agreement," November 15, 1944, folder: "Anglo-American Oil Agreement-1944," box 8, Records of the Petroleum Division, RG59.

prosecution for their participation in the decision-making process. And State had agreed to let success rest upon the prestige of the new commission members rather than legally delegated executive authority. Whether one called this a "cartel" or not would be a matter for semantic debate. In the process of negotiation, however, the real purpose had been masked with vague phrases, the scope had been changed from the Middle East to the world, and further vague phrases had been added to exclude American domestic production. But converting the document from an executive agreement to a treaty brought it under the aegis of Article VI of the Constitution, which made treaties "the supreme law of the land" along with the Constitution and laws enacted by Congress. It could now be charged, with some justification, that since "a treaty can serve as the basis for legislation . . . otherwise . . . beyond congressional competence," several clauses of the agreement could be construed as giving Congress the power to "fix the prices" and control the "production or sale of oil . . . in the United States."[102] State Department spokesmen protested that this was not the intent,[103] but the American independents, already deeply suspicious of the government as a result of the pipeline debate, reacted to this point with vehement opposition to the entire agreement.

The assistant counsel of the Special Petroleum Committee, Henry S. Fraser, advised Senator Connally and the Foreign Relations Committee of the constitutional point on August 14,[104] and George Hill of the IPAA used Fraser's argument as the keystone of a long critique presented to the PIWC on September 13.[105] Two days later Senator Moore read into the *Congressional Record* the

[102] *Congressional Record, Appendix*, 78th Congress, 2d sess., vol. 90, pt. 11, p. A4066.

[103] Letters, O'Mahoney to Hull, September 13, 1944, 800.6363/9-1344, and Hull to O'Mahoney, October 19, 1944, 800.6363/9-1344, RG59.

[104] Memorandum of conversation, Long, Rayner, and Connally et al., August 16, 1944, 800.363/8-1644, RG59. Although not so identified in the sources, circumstantial evidence is compelling that the letter referred to in Long's memorandum is the same as the one quoted by Hill in the statement cited in note 102.

[105] Minutes of meeting, Petroleum Industry War Council, September 13, 1944, folder: "Petroleum Industry War Council," box 688, item 15, RG253.

full text of Hill's statement to the PIWC.[106] The October meeting of the PIWC was a stormy one.[107] A resolution condemning the agreement had been drafted by the National Oil Policy Committee, now headed by Alvin Jacobsen after the unexpected death of John Brown. Just before the resolution was submitted, however, Ickes asked the PIWC to give him a list of proposed changes rather than simply go on record opposing ratification. A majority of the National Oil Policy Committee agreed to this approach, but Ralph Zook introduced a minority report flatly opposing ratification. Zook's resolution was defeated 21 to 11, but the National Oil Policy Committee was charged by the PIWC with drafting a proposal for a revised agreement. On November 25, the Department of Justice advised State that, after careful review, it concurred in the opinion that the agreement as a treaty did contain clauses that could be construed as giving Congress and the commission regulatory powers not originally intended, and expressed concern that it might also confer antitrust immunity on companies that cooperated with the commission.[108] Support from the majors now began to waver because, as a contemporary State Department analysis noted, the advantages to them in the proposed treaty were "not sufficiently important or definite to justify an overt conflict with the independents," the possibility that industry personnel could participate in the proposed commission was now in doubt, and it had become apparent that the agreement could probably not be "ratified in such a form as to . . . [modify] . . . the Anti Trust Laws."[109]

The December meeting of the PIWC was an impressive gathering, attended by Judge Manley O. Hudson of the Permanent

[106] *Congressional Record, Appendix*, 78th Congress, 2d sess., vol. 90, pt. 11, pp. A4065-72.

[107] Minutes of meeting, Petroleum Industry War Council, October 25, 1944, folder: "Petroleum Industry War Council," box 688, item 15, RG253.

[108] Memorandum of conversation, Rayner and Edward Levi (Antitrust Division, Department of Justice) et al., October 27, 1944, 800.6363/10-2744, and letter, Attorney General Francis Biddle to Hull, November 25, 1944, 800.6363/11-2544, RG59.

[109] Memorandum, Loftus to Rayner, October 30, 1944, folder: "Anglo-American Oil Agreement-1944," box 8, Records of the Petroleum Division, RG59.

Court of International Justice and a bevy of corporate attorneys who had worked with the National Oil Policy Committee on revisions to the agreement, and a government delegation consisting of Ickes, Davies, Acheson, Rayner, and representatives of the army, navy and PAW. Before this audience, the PIWC overwhelmingly endorsed revisions to the agreement that would, if adopted, radically curtail the powers of the binational commission and reduce it to a purely statistical and advisory function.[110] But the coup de grâce came on December 2, when Tom Connally, the senator from Texas and chairman of the Foreign Relations Committee, publicly called the agreement "unfair to the American oil industry" and announced that he was "opposed to ratification."[111] Faced with all of this opposition, the Department of State decided that it had no alternative but to withdraw the agreement for reconsideration and renegotiation. This it did on January 10, 1945.[112] The idea of an Anglo-American petroleum agreement lived on for another two and a half years, but it had been thoroughly gutted by the events of late 1944. By January 1945, the possibility that the strategic problem could be solved by a government-sponsored cartel to manage Middle Eastern production was for all practical purposes a dead issue. Other means would have to be found to provide markets for Saudi oil and an increase in Middle East production, and the Departments of State and the Navy began to shift their attention to encouraging a solution by private initiative.

Beneath the surface of the debate over the PRC and the Anglo-

[110] Minutes of meeting, Petroleum Industry War Council, December 6, 1944, folder: "Petroleum Industry War Council," box 688, item 15, RG253. The most significant defector was Max Thornburg himself, who now wanted a commission with greatly reduced powers; memorandum of conversation, Loftus and Thornburg, September 14, 1944, folder: "Memos re: Petroleum, 1943-1944," box 1, Records of the Petroleum Division, RG59; and letter Thornburg to Acheson, November 22, 1944, folder: "Assistant Secretary, Correspondence, 1941-45," box 27, Dean Acheson MSS.

[111] *New York Times*, December 3, 1944, as quoted in Shwadran, p. 330.

[112] Memorandum of conversation, Acheson, Rayner, and Connally, December 26, 1944, 800.6363/12-2644, RG59; *Congressional Record*, 79th Congress, 2d sess., January 10 and 15, 1945, vol. 91, pt. 1, pp. 179-80, 259.

American Agreement there lay the commercial issue between Socal and The Texas Company, and Jersey and Socony-Vacuum, as to whether they would compete or cooperate in the marketing of Saudi oil. Had Socal and Texas been able to secure government financing for the refinery and the pipeline, they might have been able to go it alone. But Jersey and Socony blocked the first and cooperated with the independents in killing the second. Had the Anglo-American Agreement been ratified, it could have provided Jersey and Socony with a mechanism to free themselves of the Red Line Agreement and to develop a cooperative marketing plan free of anti-trust restraints. When this approach was blocked by the independents it meant that some new plan would have to be devised to solve the postwar marketing problem of Socal and Texas and the postwar supply problem of Jersey and Socony. As matters turned out, the solution to this commercial problem dovetailed nicely with the solution to State and the Navy's strategic problem.

IV

FIELD PERSONNEL

WHILE all of this argument over oil policy had been under way in Washington, a new entity had emerged in Saudi Arabia and begun to take its place as an interest to be reckoned with in the decision-making process. The team of Socal geologists, petroleum engineers, and drilling crews that had originally opened up the concession had evolved during the war years into the Arabian American Oil Company—a highly production-oriented organization with a talent for operating in a radically different cultural environment than that of the United States. 'Abd al-'Aziz' deep interest in a larger income and modernization within the framework of an Islamic state ran in exact parallel with the interests and proclivities of the Aramco team, and the king did all he could to ensure a political environment in which Aramco's production crews could grow and flourish. It is not surprising, then, to find that Aramco personnel became the chief and most effective advocates within the inner circles of the parent corporations and the American government for increased production, cultural accommodation, and a fair share of the profits for 'Abd al-'Aziz. The story of how this came about is an interesting one.[1]

[1] This account of Aramco's early days is based on interviews with Thomas C. Barger (in Casoc/Aramco's government relations during World War II and later Aramco board chairman), in La Jolla, California, August 26, 1977; Floyd W. Ohliger (resident manager of Casoc/Aramco in Saudi Arabia during World War II), in Pineville, Pennsylvania, July 21, 1977; R. Gwin Follis (assistant to H. C. Collier in Middle Eastern matters in the 1940s and later president and board chairman of Socal), in San Francisco, August 24, 1977; Ambassador James S. Moose, Jr. (American chargé and minister resident in Jidda from 1942 through 1944), in Washington, Kentucky, April 21, 1977; Ambassador Parker T. Hart (in Dhahran as American consul from 1944 through 1946 and as consul general from 1949 through 1951), in Washington, D.C., July 15, 1977; and William E. Mulligan (in Aramco's postwar government relations organization, and unofficial

Until 1940, the "California Arabian Standard Oil Company" had existed only as a name on legal documents. For all practical purposes, the Saudi operation was an extension of the production department of Standard of California in San Francisco, headed for years by Socal vice-president Reginald C. Stoner.[2] In 1940, with commercial production assured, Socal and The Texas Company decided to formalize the organization, but from 1940 until they were moved to New York in 1949, corporate headquarters were located at 200 Bush Street in San Francisco.[3] Day-to-day decisions continued to be closely connected with Socal personnel and Socal thinking during that period.[4] This connection changed somewhat with the move to New York and even more when corporate headquarters moved to Dhahran in the 1950s. The original president of Casoc was Fred A. Davies, the Socal geologist who had conducted the initial survey of Bahrain. Davies officially reported to the Casoc board of directors (made up of representatives of the parent companies), but his office was in San Francisco and he continued to defer to Stoner on many issues. Davies served in that capacity until 1947, when he temporarily reverted to executive vice-president of Aramco in order that William F. Moore of The Texas Company might move in as president, but he moved up again later to serve as board chairman.[5] The original secretary

company historian), by telephone to New Boston, New Hampshire, December 16, 1977. Additional factual data has been drawn from the *Aramco Handbook* (1968 edition); Roy Lebkicher, *Aramco and World Oil*; an earlier version of the *Aramco Handbook*; and Wallace [E.] Stegner, *Discovery: The Search for Arabian Oil*. Attempts to obtain access to Socal and Aramco correspondence beyond what had been subpoenaed by the Senate and Federal Trade Commission were unsuccessful. Requests were turned aside with great politeness, assurances that very little remained in the records from that period, and suggestions that interviews with retired personnel would prove even more useful.

[2] Ohliger interview.   [3] Lebkicher, p. 42.
[4] Ohliger interview; Barger interview; Follis interview.
[5] Lebkicher, p. 42. The American consulate at Dhahran reported that Moore was given the presidency in 1947 to satisfy a demand by Rodgers for a larger Texas voice in the Aramco operation. That left Collier (Socal) as chairman of the Aramco board, Moore (Texas) as president, Davies (Socal) as executive vice-president, McPherson (Socal) as resident administrative officer in Saudi Arabia, and T. V. Stapleton (Texas) as general manager under him. Personal loyalties

of Casoc was James McPherson, an energetic Scotsman who had
risen through Socal's financial ranks, and who went out to Saudi
Arabia in 1944 to serve as vice-president and resident administrative officer. He resigned over policy differences in 1949 and
accepted a position with Ralph Davies' Aminol operation in Kuwait.[6] The original vice-president for production was James Terry
Duce, a dynamic, English-born geologist and intellectual-in-business from the Texas Company. Duce left briefly in late 1941
through mid-1943 to serve as director of foreign operations for
the PAW, and then returned as vice-president for government
relations in Washington until his retirement in 1959.[7] As it turned
out, the most ardent internal spokesmen for the emerging Aramco
viewpoint in the 1940s were Stoner in San Francisco, McPherson
in Saudi Arabia, and Duce in Washington.

The field personnel who remained in Saudi Arabia after the
outbreak of war were a special breed. From early 1937 until late
1944 the senior company field representative was Floyd W. Ohliger, a tough-minded Pennsylvania petroleum engineer from Socal
who had an intuitive knack for working well with 'Abd al-'Aziz
and his representatives.[8] When McPherson arrived in 1944, Ohliger continued for a time as general manager and then moved into
other assignments in Dhahran. The prewar peak of Casoc employees in Saudi Arabia was in February 1940, when there were
371 American employees, 38 American wives, 16 children, and
3,300 Arab and other non-American employees. As the European
war grew more serious, the diversion of Allied shipping to sustain
Britain and a small Italian air attack on Bahrain from Eritrea in

tended to lie with colleagues from the parent organization, and McPherson was
especially unhappy with being caught between Moore and Stapleton on matters
of internal policy; letter, Francis E. Meloy, Jr. (Dhahran), to Secretary of State,
December 12, 1948, 890F.6363/12-1248; RG59. This tension may have been a
contributing factor to McPherson's resignation two years later.

   [6] Letter, Parker T. Hart (consul general at Dhahran) to Secretary of State, July
2, 1949, U.S. Congress, Senate, Committee on Foreign Relations, Subcommittee
on Multinational Corporations, *Multinational Corporations and United States
Foreign Policy, Hearings*, pt. 7, pp. 85-89.

   [7] Letter to the author from Mrs. Ivy O. Duce, October 25, 1977.

   [8] Ohliger interview.

October 1940 drove home the fact that the Persian Gulf was vulnerable. Saudi exploratory and drilling operations were gradually shut down for lack of supplies; dependents and nonessential personnel were sent home; and operations were restricted (until mid-1944) to producing and shipping some 12,000 to 14,000 barrels of oil per day to Bahrain. The number of Americans in Saudi Arabia dropped to 111 in 1941, 87 in 1942, and 116 in 1943. The early war years therefore became known as "The Time of the Hundred Men," and the experiences shared by these twentieth-century frontiersmen had a profound effect on Aramco's style of operations for years to come.[9]

The stories about the period are legion, but none illustrates the pattern quite so well as that of the al-Kharj irrigation project. Al-Kharj is a district with large spring-fed pools fifty-five miles southeast of al-Riyad, where Finance Minister 'Abd Allah Al Sulaiman had tried to build a model farm in the late 1930s with the help of Iraqi and Egyptian technicians.[10] These early efforts were not particularly successful, and in 1941 the Saudi government asked Casoc and the American government to help. An initial geological survey was made by Thomas C. Barger, a young North Dakotan whose early field acquisition of Arabic moved him from the technical side of the business to government relations that same year, and who went on later to become president and then chairman of the Aramco board.[11] The American government lent a hand through the previously mentioned agricultural mission. The mission, which spent seven months in Saudi Arabia in 1942, included Karl Twitchell and two Department of the Interior and Department of Agriculture experts on irrigation and agriculture in the American West.[12] Then, in 1942 and 1943, the Hundred Men assisted with the acquisition of pumps and other scarce equipment from

[9] Lebkicher, pp. 41 and 55; Stegner, pp. 145-53; Barger interview.

[10] George A. Lipsky and others, *Saudi Arabia: Its People, Its Society, Its Culture*, p. 215.

[11] Stegner, p. 170; Barger interview; Mulligan interview.

[12] Letters, Welles to Roosevelt, February 12, 1942, and Berle to Twitchell, March 19, 1942, and State Department Press Release, March 25, 1942, *FR(1942)*, V, pp. 562-66; Twitchell, pp. 42-45, 165-72, 209-12.

the Middle East Supply Centre in Cairo, surveying the irrigation canal routes, and installing diesel pumping equipment. Finally, when wartime transport of the resulting produce proved a real problem, Casoc organized and operated a truck line for that purpose between al-Kharj and al-Riyad.[13] The underlying reasons for all of this were a desire for modernization on the part of 'Abd al-'Aziz, the fact that Casoc was the most convenient place to turn for technical assistance, a desire on the part of Casoc to keep on good terms with the host government, and a missionary enthusiasm for spreading American expertise on the part of the Hundred Men. A similar spirit lay behind a wide variety of educational, medical, and technical programs begun during the war years and greatly expanded in the postwar period.

None of this close collaboration would have been possible had 'Abd al-'Aziz not had a strong prior interest in modernization and great skill in doing this within the cultural constraints of Islam.[14] In the case of the al-Kharj project, the king came under extended criticism from an "unreconstructed" Wahhabi leader for "selling land" to unbelievers. He said that he knew this because he had seen Americans "take over and cultivate land at al-Kharj, . . . employ and discharge Arab workmen . . . build canals and use the precious water as they please[d]." The king summoned his critic to al-Riyad, sent a car to bring him there, and arranged a convocation of princes, courtiers, and Ulema (the authorities on Islamic law, the Shari'a), to sit in judgment. When the critic had

[13] Lebkicher, p. 43; Stegner, p. 170; Mulligan interview.

[14] The author is indebted to Professor George Rentz, a veteran Arabist and former member of Aramco's research staff in Dhahran, for pointing out the key role played by 'Abd al-'Aziz in this process. Published Aramco materials take justifiable pride in the company's role but leave the impression that modernization was something initiated by Aramco. A more balanced interpretation would view Aramco as a highly competent organization that supported 'Abd al-'Aziz in *his* efforts to modernize the country. An example is the company training program including sixty college scholarships for young Arabs in the United States, described in the *Aramco Handbook* (pp. 155-59). The program is an excellent one, but the discussion omits mention of the fact that as early as 1935 the Saudi government itself had 705 students studying abroad on scholarships, 388 in Egypt, 259 in Syria, 46 in America, and the rest in England and elsewhere; H[arry] St. John B[ridger] Philby, *Sa'udi Arabia*, p. 327.

fearlessly repeated his full accusation to the king's face, 'Abd al-'Aziz left his place, stood beside him as a fellow Muslim before the Ulema, and presented his case with great eloquence. Demonstrating a thorough knowledge of the life of Muhammad, he argued that he was merely following in the Prophet's footsteps by employing foreign experts to benefit the people of Allah and was therefore not guilty of violating Islamic law. After due deliberation, the judges returned a verdict of not guilty. The king then resumed his place. When the critic accepted the decision in poor grace, 'Abd al-'Aziz gave him twenty-four hours to apologize for his irreligious attitude or risk severe punishment, and had him led off "to meditate on his immediate future." Later that day, the king received him in private, explained quietly how his attitude had "brought discredit upon their common religion and public order," and won him over. The repentant critic was given presents and escorted in honor to his home, where he resumed his role as a loyal subject.[15]

The Hundred Men, of course, were equally sensitive to the problem of Islamic law—especially the prohibition on alcohol and the tradition of stern retribution against one who brought bodily harm to another. Among other things, the company was concerned about what might happen if an American employee, arrested and jailed for an offense under Islamic law, were to be forcibly freed by other Americans. To avoid such problems, the company adopted a policy of immediately discharging and shipping out of the country by night barge to Bahrain any employee guilty of drunken behavior in front of an Arab, physically attacking an Arab, or entering an Arab community without authorization.[16] This procedure was tacitly approved by the Saudi authorities as long as the offense was not too serious, and the first trial of an American before an Arab religious court did not occur until September 1944, when an employee was tried for accidentally killing an Arab boy while driving a company truck. (The case was settled

[15] Letter, William A. Eddy (at Jidda) to Secretary of State, December 4, 1944, 890F.001 Abdul Aziz/12-444, RG59.

[16] Letter, Hart to Secretary of State, December 20, 1944, 890F.00/12-2044, RG59; Stegner, pp. 137-138; Ohliger interview.

by payment of "blood money" to the boy's father.)[17] This pattern
was firmly established by the time newly hired construction workers began to arrive to build the Ras Tanura refinery, and the cadre
who had learned to work with the Saudis during the early war
years ran as taut a ship as possible. With slight exaggeration, Tom
Barger recalled years later that the company used to ship construction workers back to the States "by the planeload."[18] In
short, both sides did their part in reducing cultural conflict and
letting the main work of getting production under way go forward
with maximum speed.

   This, of course, was the central motif of the entire field operation—production. These were production men—geologists, petroleum engineers, and drilling crews—sent to Saudi Arabia to
organize and operate an exploration and production operation
commensurate with the vast reserves they now knew to be there.
But their enthusiasm outran the technical mission itself. As Wallace Stegner put it, "they had a faith in production, the belief that
production in and of itself is a good from which all other goods
can flow, that from production come economic security and a high
standard of living, the uncomplicated conviction that old Salim,
the Awamiri, who never in his life needed more than a camel,
a woman, and a smooth piece of sand, would be a happier man
for having learned to also like oranges, castor oil, and a Ford
pickup . . . it was a natural assumption—in the 1940s there was
little need for a commission on national goals—and the Saudis
certainly didn't seem to object."[19] There were a few observers—
such as H. St. John Philby—who were skeptical of the effect all
of this would have on Saudi society,[20] but 'Abd al-'Aziz found

   [17] Letter, Hart (to Secretary of State), October 15, 1944, file: "Confidential
Correspondence, 1944," box 683, Post files, Dhahran, Record Group 84, National
Archives Branch, Suitland, Maryland (hereafter cited as RG84).
   [18] Barger interview. In early 1945 Aramco worked out an arrangement with the
American consulate in Dhahran whereby all such dischargees would have their
passports stamped "valid only for immediate return to the United States"; memorandum, Aramco, Dhahran, to Aramco, San Francisco, February 4, 1945, file:
"Confidential Correspondence, 1/1/45-12/31/45," box 683, RG84. Both the company and the consulate wanted the dischargees completely out of the area.
   [19] Stegner, p. 168.   [20] Philby, *Sa'udi Arabia*, pp. 341-44.

the enthusiasm for production and technical modernization completely consistent with his own interests. It was a happy marriage, but it led to strong opposition from the field to any measure proposed by the parent corporations that would restrict or curtail production in any way. This viewpoint was overridden by the parents in 1946 and 1949, but—as will be seen—it had a significant impact in 1950.

Symbolic of the relationship that had begun to emerge was the new name adopted for the field operation in January 1944. During the negotiations with Ickes over government participation in the California Arabian Oil Company, Socal and Texas decided that a name change would be appropriate at the time the deal was consummated. The original plan was to change the name to American Arabian Oil Company, in exact parallel with the British precedent in Anglo-Iranian. Herbert Feis, however, suggested that such a name might convey unpleasant connotations to 'Abd al-'Aziz,[21] and the companies decided to reverse the order and make it Arabian American Oil Company. They liked this name so well that they went ahead with the change even though Ickes terminated negotiations for government participation. The decision was a fortuitous one, for 'Abd al-'Aziz increasingly came to look upon the Arabian American Oil Company as an instrument for the achievement of *his* purposes in Saudi Arabia. As early as 1944, he began to press for the transfer of Aramco's corporate headquarters to Saudi Arabia so that his government could have "easy and frequent access to the highest authorities in the company."[22] His demand was not met until the 1950s, but increasingly Aramco developed into a channel for the conveyance of Saudi interests to the inner circles of the parent corporations and the American government, as well as a highly profitable source of supply for the parent corporations. The central objective for both the company and 'Abd al-'Aziz in this period continued to be increased and uninterrupted production.

*

[21] Memorandum, Feis to Hull, September 25, 1943, 890F.6363/70, RG59.
[22] Moose interview.

THE FIELD OPERATION itself grew rapidly from late 1944 onward, as can be seen from Table IV-1. The company had long recognized the need for refinery capacity in addition to that located on Bahrain, in order to convert high sulphur Saudi oil into products easier to sell in the world market. A 3,000 barrels/day refinery had been built at Ras Tanura in the autumn of 1940 to supply company and government needs within Saudi Arabia, and plans had been completed for a 25,000 barrels/day refinery at the same location when the war intervened. Both the small refinery and work on the larger refinery were closed down in 1941.[23] In early 1943 the Foreign Operating Division of the Petroleum Administration for War began working with the company on plans for a 100,000 barrels/day refinery at Ras Tanura as part of the wartime supply effort. It was possible government funding for this project that Ickes used as bait to lure the companies into selling the government an interest in the Saudi concession. When that project collapsed, Socal and Texas decided to go ahead on their own with a 50,000 barrels/day refinery, which was more consistent with their projected postwar requirements.[24]

The project was a massive one. It involved not only the refinery itself, but new storage tanks, new loading lines, a new T-shaped pier and wharf for docking tankers, and a submarine pipeline to Bahrain, where the Bapco refinery was also being enlarged. Most of the construction workers were provided by the International Bechtel McCone Company, and the government assigned high military priorities to the steel and other critical materials required. Shipping space was also given high priority, and despite the delay encountered by the loss of one freighter en route, construction work was in full swing by the end of 1944. The submarine pipeline to Bahrain was completed and placed in operation in April 1945, and the Ras Tanura refinery went into partial operation in Sep-

[23] Lebkicher, p. 41; Barger interview.
[24] Letter, Snodgrass to Carter, May 12, 1943, and letter Davies (Casoc) to Ickes, September 6, 1943, file: "Refinery File"; and "Report on . . . audit and investigation . . . ," January 31, 1945, file. "Audit Report"; Records Regarding Claim Against PRC by Arabian American Oil Company, RG234.

TABLE IV-1
ARAMCO EMPLOYMENT IN SAUDI ARABIA, 1938-1950

At the end of	Americans	Saudi Arabs	Others	Total
1938	236	2,745	104	3,085
1939	322	3,178	141	3,641
1940	226	2,668	156	3,050
1941	111	1,647	95	1,853
1942	87	1,654	84	1,825
1943	116	2,692	74	2,882
1944	961	7,585	514	9,060
1945	775	7,500	1,665	9,940
1946	1,158	7,297	636	9,091
1947	1,855	12,018	2,374	16,247
1948	2,586	12,226	3,844	18,656
1949	2,296	10,063	3,145	15,504
1950	2,353	10,767	3,742	16,862

SOURCE: Lebkicher, p. 55.

tember and full operation in December 1945.[25] Output for the next five years ran as shown in Table IV-2.

The second requirement for improving the marketability of Saudi oil was a pipeline from the Persian Gulf to the Mediterranean. The companies began preliminary planning for this in 1943,[26] and it was this project that Ickes seized upon in early

TABLE IV-2
RAS TANURA REFINERY OUTPUT, 1946-1950

Year	Barrels
1946	29,297,101
1947	39,065,010
1948	45,086,139
1949	46,269,619
1950	38,364,335

SOURCE: Lebkicher, p. 46.

[25] Lebkicher, pp. 45-46.
[26] Memorandum, Office of the Petroleum Advisor to Murray et al., December

1944 as an alternative device for establishing a governmental presence in the Middle East. When federal funding for the pipeline failed to materialize because of opposition from other American companies, Socal and Texas went ahead with their planning and began serious internal discussions on how to raise the money. The distance to be covered was approximately 1,100 miles, equal to the air distance from Miami to New York or Vancouver, B.C. to Los Angeles.[27] Most of that was along the northern border of Saudi Arabia itself, but the final portion would go through Transjordan, and then through either Palestine or Syria and Lebanon. The more direct route would have been through Palestine, but in view of the developing Palestinian issue, the company deferred a final decision on which route to use. Negotiations for transit rights through Transjordan and Palestine were begun by William J. Lenahan with British mandate authorities in Jerusalem in late 1944 and transferred to London in March 1945.[28] In July, Socal and Texas organized a separate joint subsidiary—the Trans-Arabian Pipe Line Company (Tapline), a Delaware corporation—to handle the project.[29] On January 7, 1946, Lenahan signed a transit agreement with the British high commissioner for Palestine in Jerusalem,[30] but an agreement with Transjordan was delayed by the ending of the British mandate there in early 1946, and negotiations were continued with the new, independent Jordanian government. The rest of the Tapline story will be taken up in Chapter VI, but at this point it is significant to note that the

---

4, 1943, 800.6363/1397-1/2, RG59. The memorandum reported a discussion with Thornburg, who was preparing a report on the pipeline project for Socal and Texas. See also, letter, Davies (Socal) to Murray, December 27, 1943, and letter, Rayner to Davies, January 7, 1944, *FR(1944)*, V, 8-9, and 12-13.

[27] Lebkicher, p. 50.

[28] Telegram, Hamilton (Aramco, London) to Duce (Aramco, Washington), March 28, 1945, copy in 867N.6363/3-2745, RG59; letter, Duce to Richard Sanger (Division of Near Eastern Affairs), November 23, 1945, 867N.6363/11-2345, RG59; and telegram, Winant to Secretary of State, November 27, 1945, *FR(1945)*, VIII, 60-61.

[29] *Aramco Handbook*, p. 148.

[30] The full text of the Palestine transit agreement is in the British archives, V52598, E1419/1419/31, FO/371.

question of financing for the pipeline had not been resolved by late 1945.

The most important aspect of Aramco's postwar performance was increased production, and, as can be seen from Table IV-3, the increase began in 1944 and continued steadily thereafter. With the low cost of production in Saudi Arabia, this proved to be a profitable enterprise for both the company and the king, but just how profitable is difficult to assess. An estimate of direct production costs for January through July 1945, provided by James Terry Duce to a Senate committee in 1948, gave a cost per barrel of:[31]

Operating expenses:

Royalty	$.2199
Other	.0796
General expenses	.0315
Depletion and depreciation	.0750
	$.4060

This did not include such things as interest on money borrowed, general survey expenses, or about $.25 per barrel that Rodgers told the committee was a fair estimate to cover losses incurred by the parent companies in unprofitable exploration all over the world.[32] Whatever the realistic cost per barrel, it was substantially below the f.o.b. Ras Tanura price of $.85 to $1.05 for which Aramco was selling fuel oil to France and the U.S. Navy respectively during that same period.[33] The profit margin for Saudi oil, with its comparatively low production costs, was a substantial one.

Exactly what this translated into in terms of corporate profits is even more difficult to determine. For one thing, Aramco was not publicly owned, and its earnings were therefore not published

[31] U.S. Congress, *Petroleum Arrangements with Saudi Arabia*, pp. 25008 and 25388.

[32] Ibid., pp. 24914 and 25009.

[33] Ibid., pp. 24950-51. Prices varied with the interest in breaking into a new market; they ranged as low as $.70 per barrel for Japan and as high as $1.21 per barrel for UNRAA in Italy during that same period.

TABLE IV-3
ARAMCO CRUDE OIL PRODUCTION, 1938-1950

Year	Barrels Daily	Barrels Annually
1938	1,357	495,135
1939	10,778	3,933,903
1940	13,866	5,074,838
1941	11,809	4,310,110
1942	12,412	4,530,492
1943	13,337	4,868,184
1944	21,296	7,794,420
1945	58,386	21,310,996
1946	164,229	59,943,766
1947	246,169	89,851,646
1948	390,309	142,852,989
1949	476,736	174,008,629
1950	546,703	199,546,638

SOURCE: *Aramco Handbook*, p. 135.

as a discrete item separate from the consolidated statements of its parents. Secondly, the profits attributed to Aramco were really a matter of bookkeeping and could vary with such things as the price used for transfers to other components of the same corporate network. If oil were "sold" to Caltex or Tapline at a discount, for example, this would decrease Aramco's "profit." Which component of the network of subsidiaries showed a "profit" frequently made a difference to the various parent companies, and to the Saudi government for tax purposes after 1950. For these reasons, there were recurring debates within the corporate network and with the Saudi government on the subject of transfer prices. As a result, Aramco's book "profits" reflected in part the outcome of these debates as well as actual operating experience. All of this notwithstanding, Professor Zuhayr Mikdashi of the American University of Beirut has made a careful study of documents in the public domain and has arrived at estimates of Aramco's net book profits (see Table IV-4).[34] Even if these figures are not

[34] Zuhayr Mikdashi, *A Financial Analysis of Middle Eastern Oil Concessions: 1901-65*, pp. 115-20. Mikdashi also served as an early advisor to OPEC and to the government of Kuwait.

TABLE IV-4
Aramco Payments to the Saudi Government, Net Profits, and
Dividends, 1939-1950
(In millions of dollars)

Year	Payments to the Saudi Government	Net Book Profits After Taxes	Dividends to Parents
1939	3.2[1]	.7[3]	None
1940	2.5	.3	None
1941	2.0	.2	None
1942	2.0	.2	None
1943	2.0	.1	None
1944	2.5	2.8	None
1945	5.0	3.6	None
1946	12.5	23.0	None
1947	17.5	49.4	37.1[4]
1948	50.8[2]	80.3	25.0
1949	39.0	115.1	61.7
1950	56.7	127.4	87.9

Sources: *Payments to the Saudi Government*: Mikdashi, pp. 122 and 184. His sources were Saudi Arabia, *Arbitration*, vol. 1, p. 15, chart 8 (1939-49), and Saudi Arabian Monetary Agency, *Annual Report 1383-84 A.H.* (July 18, 1965, p. 42 (1950). *Net Profits and Dividends*: Mikdashi, p. 120 (1939-49). His sources for net profits were Socal and Texaco annual reports (1939-45); Socal annual reports to the Security and Exchange Commission, Washington, D.C. (1942-43 and 1947); U.S. Congress, *Petroleum Arrangements with Saudi Arabia*, p. 24901 (1946); and calculations based on Socal's 30% interest in Aramco and data in *Moody's Industrials, 1955*, p. 2684. His sources for dividends were calculations based on Socal's 30% interest in Aramco, and data in Socal's *Annual Report*, 1947, pp. 16 and 19 (1947), and *Moody's Industrials, 1955*, p. 2683 (1948-49). Sources for net profits and dividends for 1950 were the author's own calculations based on Socal's 30% interest in Aramco and data in *Moody's Industrials, 1955*, pp. 2683-84.

1. Includes $1.16 million paid in accordance with the Supplemental Concession Agreement of May 31, 1939.
2. Includes $19.32 million paid in settlement of the gold pound controversy.
3. 1939 was the first year in which Casoc/Aramco showed a net profit.
4. 1947 was the first year in which dividends were paid out to the parent companies.

precisely correct, they appear to have been constructed with considerable care, and they clearly demonstrate that Aramco was a profitable enterprise.[35] If taken alone, however, the data on profits can be a bit misleading with regard to the parent companies' financial position at the end of 1946. Up to that point they had advanced Aramco approximately $80 million without repayment, and they were faced with financing a pipeline at an estimated cost of well over $100 million.[36] The future looked bright, but the amount of risk capital required was quite high. The profit figures are especially interesting when compared with data on payments made to the Saudi government during the same period, as published in the records of a 1955 arbitration between Aramco and the Saudi government.[37] Until 1950, these payments were fixed by the original concession agreement at 4 shillings (gold) per ton, and as sales picked up, earnings for the Saudi government began to lag far behind those of the company. This discrepancy caused

[35] An even more interesting figure would be a realistic estimate of Aramco's rate of return on investment, but this is even more elusive than "profits." For 1956 through 1960 Mikdashi (p. 182) reports rates estimated by OPEC for Aramco at 57 to 71 percent, with an average close to 61 percent. These figures may be high, and for reference they may be compared with an estimate of the overall rate of return for all United States direct investment in foreign petroleum reported by an analyst more sympathetic to the industry—Neil H. Jacoby, *Multinational Oil: A Study in Industrial Dynamics* (p. 248). For the same period (1956-60) Jacoby gives an overall rate of return of 28.8 percent to 13.6 percent, with an average of 19.4 percent. Mikdashi (p. 183) suggests that at least after 1950, many companies operating in the Eastern Hemisphere posted most of their book profits at the production stage, leaving a much lower rate of return in other phases, and an across-the-board average somewhere in between. All of these complications notwithstanding, it is clear that Aramco was a highly profitable investment.

[36] Rodgers testified in 1948 that as of January 1, 1948, Socal had invested $39,883,000 in Aramco and The Texas Company had matched that amount. In addition, Texas had paid Socal $21 million for its 50 percent interest. Rodgers argued that this was additional investment, which was true from Texas' viewpoint, but it was not money required to be repaid to the parents and has not been counted in arriving at the figure of $80 million. U.S. Congress, *Petroleum Arrangements with Saudi Arabia*, pp. 24900-902.

[37] Saudi Arabia, *Arbitration Between the Government of Saudi Arabia and Arabian American Oil Company*, vol. 5, [First] *Memorial of Arabian American Oil Company*, p. 15, chart 8.

'Abd al-'Aziz in 1949 and 1950 to press for a larger share of the revenue, a move that led to the fifty-fifty profit sharing negotiations discussed in Chapter VI.

*

TO SUM UP, by the end of World War II the original Socal field operation had evolved into the Arabian American Oil Company, a highly competent, self-reliant and increasingly assertive production organization that got along exceptionally well with 'Abd al-'Aziz. In addition to producing oil for the benefit of both the parent companies and the king, Aramco had become a contributor to the modernization of Saudi Arabia and a spokesman for Saudi interests with the parent companies and the American government.

V

AND CORPORATE OFFICERS

RETURNING to Washington and January of 1945, we find that withdrawal of the Anglo-American Agreement from the Senate set in motion a major realignment of positions and a full-scale rethinking of policy toward Saudi oil. The objectives of the Departments of State and the Navy remained unchanged and firm as ever: Middle Eastern production in general and Saudi production in particular should be increased as rapidly as possible to decrease the drain on Western Hemisphere reserves, and all necessary measures should be taken to keep the Saudi concession firmly in American hands. But, as it became increasingly apparent that a revised Anglo-American Agreement would be too weak to accomplish those objectives, other means were sought out. Ickes was allowed to renegotiate a now meaningless agreement, and State and the Navy turned to airfield construction and Export-Import Bank loans for short-term aid to 'Abd al-'Aziz, and encouragement of private initiative to solve the longer-term marketing problem. The most significant byproduct of the 1945 debate appears to have been strong governmental endorsement of a decision by Socal and Texas to invite Jersey and Socony-Vacuum to become partners in the Aramco venture, in order to add funds and markets for rapid development of the concession. The reasons for governmental endorsement become clearer, however, if we first trace the debate in the spring of 1945 over a revised agreement and aid to the Saudis.

*

AS ALREADY NOTED, the turning point for the Anglo-American Agreement occurred in late 1944, when Ickes asked the Petroleum Industry War Council to propose revisions rather than flatly condemn the entire agreement, and the strength of the industry demand for revision plus Senator Connally's opposition forced State to withdraw the agreement from the Senate for renegotiation. This development shifted effective control of the contents of the agreement to the PIWC, and Ickes quickly moved to align himself with this new power center by supporting every one of the industry demands in the intragovernmental debates that followed. The original President's Committee was reassembled on January 9 under the chairmanship of Stettinius, who had replaced Hull as secretary of state, and it directed Davies and Rayner to meet with Jacobsen's PIWC Committeee on National Oil Policy to gain a thorough understanding of the reasoning behind the industry's revised draft.[1] These meetings were held on January 15, 16, and 17, and when Rayner and Davies reported back to the President's Committee on January 20, Stettinius had left Washington for the Yalta Conference. As vice-chairman, Ickes again took charge, and he retained effective control of the process of reconsideration for the next nine months, at one point even forbidding State to retain a copy of the revised draft for internal study. In this capacity, Ickes now assumed the role of advocate for the industry position, which called for language clearly excluding domestic American production, limiting the International Commission to purely advisory functions, and providing antitrust protection to companies that elected to comply with the commission's advice. This was fine with Ickes, who appears never to have understood the role the commission was to have performed. He thought that the objective was an agreement on general principles to keep the British from encroaching on American concessions, and he appears never to have understood State's broader thinking on the subject.

In the meantime, both organizational and philosophic changes

[1] Memorandum, Darlington to Haley and Clayton, February 2, 1945, file: "Anglo-American Oil Agreement, 1945," box 4, Records of the Petroleum Division, RG59; and minutes of meeting, Petroleum Industry War Council, January 17, 1945, file: "Petroleum Industry War Council," box 688, item 15, RG253. For a clear contemporary exposition of the industry's position, see J. H. Carmical, *The Anglo-American Petroleum Pact: A Case Study in the Negotiation of Postwar Agreements*, pp. 28-36. For Ickes' thinking on this subject, see his diary entries for January 13, 21, and 27, February 24 and 25, March 1, 17, and 24, April 4 and 29, and June 2 and 9, 1945, Ickes MSS.

had taken place in the Department of State, and enthusiasm for the agreement had begun to wane. In 1944, most of the personnel of the Petroleum Division had been separated from the petroleum advisor (Rayner) and transferred to the Office of Economic Affairs under Assistant Secretary William L. Clayton.[2] Rayner continued to offer advice, but day to day decisions were increasingly made in the Office of Economic Affairs. Clayton's staff was deeply involved in postwar economic planning, and these plans emphasized free trade, multilateral as opposed to bilateral agreements, and firm opposition to anything resembling a cartel. This was not a particularly hospitable environment for the Petroleum Division's original idea, but in practical terms it meant firm department opposition to antitrust immunity for the oil companies, as incompatible with its opposition to cartels.[3]

In late January, John Loftus drafted a perceptive memorandum from his new vantage point within Clayton's organization, describing the situation that had now developed. He agreed with a conclusion reached by Rayner that because the industry had been able to force withdrawal of the original agreement, it was now clear that "no agreement could be negotiated or would be workable which did not have the support of the . . . industry." But the price of this support was so great that it was now doubtful that an agreement could be negotiated that would "further . . . [any] . . . legitimate objective of public policy" or "serve . . . [any] . . . useful purpose." The original idea, he pointed out, had been "conservation of Western Hemisphere . . . reserves . . . in the interest of national security . . . [by] . . . curtailment of the flow of petroleum . . . from Western Hemisphere sources to Eastern Hemisphere markets . . . [and] . . . expansion of

[2] Memorandum, "The Petroleum Division," by G. M. Richardson Dougall, October, 1944, pp. 6-8, file: "Petroleum-General," box 48, Harley Notter files, RG59.

[3] Memorandum, "Objections of the State Department to the Proposed Revision of the Anglo-American Petroleum Agreement," by H[oward] S. P[iguet], March 28, 1945, file: "Anglo-American Oil Agreement, 1945," box 4, Records of the Petroleum Division, RG59; memorandum, Haley to Clayton, March 19, 1945, 800.6363/3-1945, RG59; and memorandum, Joseph C. Grew (acting secretary of state) to Roosevelt, March 21, 1945, 800.6363/3-2145, RG19.

production in [the] Eastern Hemisphere.'' The original instrument for achieving this had been the International Commission, with limited, but clear, executive power. At this point, however, the industry was demanding that the agreement "exclude altogether . . . the operations of the American petroleum industry . . . [which made up] . . . approximately 65% of the [world] oil industry . . . ; be predicated [solely] upon 'voluntary compliance' . . . ; avoid all reference to criteria for the international organization of the industry that might . . . imply any inhibitions upon the American . . . industry . . . ; and provide immunity from [antitrust] prosecution . . . with respect to actions taken in direct or indirect compliance with recommendations of an international commission . . . even though such recommendations would not be binding upon petroleum companies.''[4] The inference was that an agreement strong enough to accomplish the original objective was no longer possible, and the record clearly indicates that from this point on State's support of the renegotiation process was purely pro forma. The proposed antitrust clause had to be put to rest, but aside from that, Ickes was allowed to negotiate whatever agreement he wanted. The department had to find other means for increasing Middle East production and maintaining a solid American grip on the Saudi concession.

The antitrust issue was fully joined on February 22, when, at Senator Connally's request, the PIWC's National Oil Policy Committee and representatives of the Department of State and the Petroleum Administration for War met with the Senate Foreign Relations Committee for informal discussion of the industry proposals.[5] The central issue was antitrust, and inclusion of an immunity clause was staunchly defended by Alfred Jacobsen, chairman of the National Oil Policy Committee; Judge Hudson, legal

[4] Memorandum, "The Anglo-American Oil Agreement," by J. A. L[oftus], January 31, 1945, 800.6363/1-3145, RG59.

[5] Minutes of meeting, President's Committee on Oil, February 20, 1945, 800.6363/2-2445, RG59; memorandum, Mason to Clayton, February 21, 1945, 800.6363/2-2145, RG59; minutes of meeting, Petroleum Industry War Council, March 14, 1945, file: "Petroleum Industry War Council," box 688, item 15, RG253.

counsel for the Policy Committee; Ralph Zook, president of the Independents; Colonel Frank Drake, president of Gulf; Brewster Jennings, president of Socony-Vacuum; and Eugene Holman, who had replaced Ralph Gallagher as president of Jersey in mid-1944. They argued—with some justification—that because compliance with the commission's recommendations was now voluntary, an agreement among producers to do such things as reduce Latin-American production in favor of Middle East production could expose them to antitrust prosecution unless specific immunity were granted. Senators Connally, Joseph C. O'Mahoney, and Robert M. LaFollette, Jr., expressed reservations about amending the antitrust laws by way of a treaty, and Assistant Secretary of State Acheson declined comment on the grounds that the department had not yet had access to the revised draft for thorough study.[6]

After the meeting, Ickes resorted to his old tactic of appealing to Roosevelt for support against the Department of State's "heel-dragging," and the president agreed to look into the matter as soon as he returned from Warm Springs.[7] The President, however, never returned from Warm Springs, and with Roosevelt's death Ickes lost his privileged position in the federal bureaucracy. After further discussion at the working level, State won its point, and Ickes acquiesced in deletion of the antitrust clause from the revised draft. He no longer had the leverage to do otherwise.[8] The chief

[6] Memorandum, Mason to Clayton, February 22, 1945, 800.6363/2-2145, RG59. The British Foreign Office was kept well informed of all of these developments by its embassy in Washington; telegram, Halifax to Foreign Office, February 27, 1945, V50380, W3137/12/76, FO/371; and letter, Starling to R. A. Gallop (Foreign Office), August 30, 1945, V50386, W11774/12/76.

[7] Letter, Ickes to Roosevelt, March 12, 1945, memorandum, Grew to Roosevelt, March 21, 1945, and letter, Roosevelt to Ickes, March 24, 1945, file: "Anglo-American Oil Agreement," box 115, P.S.F., Roosevelt MSS; memorandum, Ickes to Roosevelt, March 26, 1945, and memorandum, Roosevelt to Stettinius, March 27, 1945, file: "Anglo-American Oil Agreement, 1945," box 4, Records of the Petroleum Division, RG59.

[8] Letter, Ickes to Stettinius, April 12, 1945, 800.6363/4-1245, RG59; minutes of meeting, Petroleum Industry War Council, May 16, 1945, file: "Petroleum Industry War Council," box 688, item 15, RG253; minutes of meeting, President's Committee on Oil, June 6, 1945, 800.6363/6-1345, RG59; letter, Ickes et al., to

significance of this decision however, was the fact that the majors now lost interest in the agreement. They gave Ickes credit for a good try, but from this point on their support for the renegotiation process was also pro forma. The statements of general principles remaining in the draft agreement were nice, but of insufficient value to warrant a major effort to ensure ratification.

The remainder of the story of the Anglo-American Agreement is anticlimatic. The next step was renegotiation with the British, but this was delayed slightly by the British elections and change of governments in mid-1945. This time, Ickes led an American delegation to London, and talks opened September 17 with a British delegation led by the new minister of fuel and power, Emanuel Shinwell. In addition to an official delegation composed of Davies, Rayner, and six others from the Petroleum Administration for War and the Department of State, Ickes brought along an advisory group that included Jacobsen, Zook, Hill, William R. Boyd, Jr., president of the American Petroleum Institute and chairman of the Petroleum Industry War Council, W. Alton Jones, president of Cities Service, and Joseph E. Pogue, vice-president of the Chase National Bank.[9] Noticeably absent were any representatives of the major international companies. Even the Lon-

---

President Harry S. Truman, June 14, 1945, file: "President's Committee on the Anglo-American Oil Agreement," box 30, President's Secretary's file, Harry S. Truman Papers (hereafter cited as Truman MSS); and telegram, Grew to Amembassy, London, June 20, 1945, 800.6363/6-2045, RG59.

[9] Telegram, Foreign Office to Washington, July 6, 1945, V50384, W9162/12/76, FO/371; and telegram, Acheson to Amembassy, London, September 12, 1945, 800.6363/9-1245, RG59. The other official American delegates were John A. Loftus, now special assistant to the director of the Office of International Trade Policy in the Department of State, Robert E. Hardwicke, general counsel for the PAW, Gordon M. Sessions and Samuel Botsford, foreign relations advisors in the PAW, and Victor Barry, petroleum attaché, and James C. Sappington, second secretary, from the American embassy in London. In addition to Shinwell, the initial British delegation included Minister of State Philip Noel Baker, Sir Norman Duke, F. C. Starling, and K. L. Stock of the Ministry of Fuel and Power, David Waley of the Treasury, R. A. Gallop of the Foreign Office, and Victor Butler, the British Petroleum Representative in Washington; minutes of Plenary Session no. 1, Anglo-American Conversations on Petroleum, September 18, 1945, V50388, W13146/12/76, FO/371.

don *Times* commented that "the composition of Mr. Ickes party" suggested that an agreement imposing "general control of output or sales" was highly unlikely.[10]

The Times was absolutely correct. The discussions lasted less than a week, and they were devoted primarily to changes in phraseology that would limit the scope of the agreement and the powers of commission.[11] When the revised agreement (Appendix B) was signed on September 24, it restricted the commission to studies and reports, and to avoid all doubt it included a clause to the effect that "no provision in this Agreement shall be construed to require either Government to act upon any report or proposal made by the Commission, or to require the nationals of either Government to comply with any report or proposal made by the Commission." At the insistence of the American independents, the new document also provided that "nothing in this Agreement shall be construed as impairing . . . the right to enact any law or regulation, relating to the importation of petroleum into the country of either Government."[12] The British readily agreed because this change eliminated any question about their right to restrict imports to sterling oil, and it thereby resolved what had been the major problem in the 1944 negotiations.[13]

The revised agreement was transmitted to the Senate on November 1, and hearings were initially scheduled for March 4,

[10] *The Times*, September 13, 1945, as quoted in letter, Sappington to Secretary of State, September 14, 1945, 800.6363/9-1445, RG59. The British government also took note of the same point; minutes of meeting of the Ministerial Oil Committee, September 12, 1945, V50387, W12258/12/76, FO/371. Ickes' diary suggests that the trip to London and on to Paris with several members of his delegation was more of a social event than a serious working conference. It is clear that he and his party enjoyed the trip immensely; diary entries for October 6 and 7, 1945, Ickes MSS.

[11] Minutes of Plenary Sessions nos. 1 and 2, Anglo-American Conversations on Petroleum, September 18 and 20, 1945, V50388, W13146/12/76, FO/371.

[12] Great Britain, *Agreement on Petroleum Between the Government of the United Kingdom of Great Britain and Northern Ireland and the Government of the United States of America*, Command Paper no. 6683.

[13] Minutes of meeting of Ministerial Oil Committee, September 24, 1945, V50388, W12900/12/76; and Cabinet Paper C.P.(45)194, "Anglo-American Oil Agreement," September 26, 1945, V50388, W13147/12/76.

1946.[14] This time, general industry reaction was good, and resolutions of support were passed by the Independent Petroleum Association, the American Petroleum Institute, and the Petroleum Industry War Council.[15] But Harold Ickes finally retired from government service at the age of seventy-one in February 1946, and the agreement was left without ardent supporters in either government or industry. Five days after Ickes retired, an internal State Department memorandum pointed out that the agreement was now a useless "orphan" that could be killed by "having the hearings indefinitely postponed, . . . dampening the industry's enthusiasm . . . [and] . . . presenting testimony which damns it with faint praise."[16] Although there is no documentary evidence that this course of action was ever formally adopted, there is strong circumstantial evidence that it was, in fact, the course taken. At State's suggestion, the hearings were postponed indefinitely—to avoid complicating debate on the loan to Britain then being considered.[17] And in August 1946, John Loftus, now chief of the department's petroleum division, made statements in the course of a radio panel discussion that re-aroused industry suspicions that the agreement might be only the first step in a long-term plan for international control.[18] At its October meeting, the Independent Producers Association reversed itself and went over into opposition,[19] and in November, and again in January 1947, the board of directors of the American Petroleum Institute qualified its earlier endorsement with a series of reservations designed

[14] U.S. Congress, Senate, Committee on Foreign Relations, *Petroleum Agreement with Great Britain and Northern Ireland, Hearings*, June 2-25, 1947, p. 1; and memorandum, Wilcox to Clayton, February 19, 1946, 800.6363/2-1946, RG59.

[15] John W. Frey and H. Chandler Ide, *A History of the Petroleum Administration for War, 1941-1945*, pp. 286-87.

[16] Memorandum, Wilcox to Clayton, February 19, 1946, 800.6363/2-1946, RG59.

[17] Letter, Acting Secretary of State to Officer in Charge of the American Mission, London, July 3, 1945, 800.6363/5-946, RG59.

[18] U.S. Congress, *Petroleum Agreement with Great Britain*, pp. 68-69, 117-24, 251-63; telegram, Butler to Starling, August 21, 1946, V53054, UE3847/721/53, FO/371.

[19] U.S. Congress, *Petroleum Agreement with Great Britain*, p. 251.

to clarify its opposition to anything resembling federal or international control.[20]

When the Foreign Relations Committee finally heard testimony in June 1947, those industry representatives who had been most closely associated with the debate and negotiating process (Jacobsen, Hill, Zook, Boyd, Pogue, Judge Hudson, and Senator Moore) went on record favoring ratification—providing the American Petroleum Institute reservations were incorporated in the act of ratification.[21] From the majors, only Eugene Holman of Jersey testified—in support of the agreement as "desirable" but not "necessary."[22] The opposition came primarily from Texas and the independents, and included statements or testimony from Governor Beauford H. Jester of Texas, Olin Culberson and W. J. Murray of the Texas Railroad Commission, Russell B. Brown, counsel for the Independent Petroleum Association, Henry S. Fraser, representing Sinclair Oil, H. J. Porter, president of the Texas Independent Producers and Royalty Owners Association, and a bevy of independent oil men from Texas and Colorado.[23] The opposition even introduced resolutions from the board of regents of the University of Texas and the Texas State Teachers Association denouncing the agreement.[24] Despite all of this, the Foreign Relations Committee recommended ratification in its final report to the Senate.[25] The agreement, however, was never called up for a vote, and there it languished until finally withdrawn by the Department of State as a dead issue in 1952[26]—

[20] Ibid., pp. 80-84; letter, Boyd to Truman, February 7, 1947, file: "OF56A(1945-46)," box 270, Truman MSS.

[21] U.S. Congress, *Petroleum Agreement with Great Britain*, pp. 7-21, 80-86, 135-176, 276-78.

[22] Ibid., p. 190.

[23] Ibid., pp. 72-80, 250-63, 279-92, 309-25.

[24] Ibid., pp. 281-82.

[25] U.S. Congress, Senate, Committee on Foreign Relations, *Anglo-American Oil Agreement*. The committee vote was 11 to 1 in favor, with only Senator Connally in opposition. For a clear contemporary discussion of the status of the agreement in the late 1940s, see Raymond F. Mikesell and Hollis B. Chenery, *Arabian Oil: America's Stake in the Middle East*, pp. 95-100.

[26] Editorial note in U.S. Department of State, *Foreign Relations of the United States (1945)*, VI, 244 (hereafter cited as *FR[date]*).

killed by faint praise from the few remaining proponents and intense opposition from the Texas independents.

*

IN EXACT PARALLEL with the demise of the Anglo-American Agreement, a new plan for the rescue of 'Abd al-'Aziz had begun to emerge within the Departments of State, War, and the Navy. Discussions among the Americans, British, and Saudis regarding the size and content of the 1944 supply and subsidy program had dragged on interminably,[27] and when the results were finally announced to the king in August 1944, he expressed grave disappointment.[28] Resorting to old tactics, he pointed out that "the combined supply and financial aid . . . [was] . . . less than that received from the British [alone] in 1943 . . . [and] . . . the net result . . . [was] . . . that Britain . . . [was] . . . relieved of half of her . . . obligations, without any benefit whatsoever to Arabia." He was in dire straits and concerned that America might "lose interest in his distant land, after the war," or "permit her policy to be dictated by influential pressure groups" more favorable to Palestine. As evidence of no such long-term unfriendly intent, he suggested not only an increase in the 1944 supply program, but also consideration of ways "in which Saudi Arabia and America . . . [might] . . . collaborate . . . on a basis that leads far beyond an end to the war." As added incentive, the king hinted that he would be "hospitable to a treaty for American aviation rights [which the Army Air Corps had requested in July] if thereby he could secure financial resources soon."[29]

[27] *FR(1944)*, V, 670-722.

[28] Telegrams, Moose to Secretary of State, August 6, 1944, *FR(1944)*, V, 723-25.

[29] Telegram, William E. Eddy (appointed minister to Saudi Arabia) to Secretary of State, September 7, 1944, *FR(1944)*, V, 734-36. Initial interest in "flyover" rights and construction of an airfield at Dhahran stemmed from a desire to shorten the military air route from Cairo to Karachi and to locate a field on better terrain than available at Bahrain. 'Abd al-'Aziz had referred the initial request to the British for their advice, and the British had raised a series of objections that delayed a final decision for some time; letter, Moose to Yusuf Yassin (Saudi Arabian acting minister of foreign affairs), July 29, 1944; telegram, Eddy to

In response to this overture, Wallace Murray in NEA began work on a new short-term plan to aid 'Abd al-'Aziz in the interim before oil royalties became large enough to solve his financial problems. By October 18, the department had developed a "Proposed Plan for Safeguarding American Interests in Saudi Arabia" that estimated the total Saudi budgetary deficits from June 30, 1945, through June 30, 1949, at $26.7 million, and proposed to fund this with a $10 million payment on a fifty-year airfield lease, and a $16.7 advance royalty payment through Aramco on an option contract on 100 million barrels of oil for the navy. In addition, the plan called for the War Department to undertake various roadbuilding and construction projects desired by the Saudi government.[30] Since the most immediate issue was construction of an airfield, Murray conferred first with representatives of the Army Air Corps, and it was agreed that the best approach would probably be for the War Department to request that the Department of State make recommendations regarding "methods by which the interest of the United States [in Saudi Arabia] could be safeguarded."[31] This Secretary of War Stimson did on October 27, with a copy to the secretary of the navy.[32]

Secretary of State, September 4, 1944; telegram, Hull to Winant, October 17, 1944; memorandum, "The Dhahran Airfield Matter" (November 9, 1944); *FR(1944)*, V, 661-63, 666-68.

[30] Memorandum, "Proposed Plan for Safeguarding American Interests in Saudi Arabia," October 18, 1944, file: "Saudi Arabia-Oil," box 6, Records of the Petroleum Division, RG59. The document is unsigned, and it is not clear whether it was drafted in the Petroleum Division or in the Office of Near Eastern and African Affairs. For comment on flaws in the original document, see memorandum, Loftus to Sappington, October 18, 1944, file: "Memos re: Petroleum, 1943-1944," box 1, Records of the Petroleum Division, RG59. For an excellent account of efforts within the American government to formulate a plan for aiding Saudi Arabia at the end of the war, see Miller, *Search for Security: Saudi Arabian Oil and American Foreign Policy, 1939-1949*, pp. 129-49.

[31] Memorandum, Major General J. E. Hull (assistant chief of staff) to Chief of Staff and Secretary of War, October 24, 1944, file: "091 Saudi Arabia-Miscellaneous," box 78, Record Group 353, Records of the State-War-Navy Coordinating Committee (hereafter referred to as RG353).

[32] Letter, Stimson to Secretary of State, October 27, 1944, *FR(1944)*, V, 748-51.

Stimson's letter set off extensive inter and intradepartmental consultations on the subject, and the original proposal was extensively modified. With some justification, it was concluded that direct expenditure of a sizable amount of government funds for air rights and a naval petroleum reserve in Saudi Arabia would meet with the same opposition from the Congress and competing oil interests that had killed the PRC pipeline. It was now clear that "the American producers fear[ed] that an unregulated flood of low-cost Arabian oil . . . [would] . . . both reduce the market for their own oil, and reduce the price of such oil as they . . . [could] . . . still market." For this reason, governmental actions in regard to Arabian oil would be "the object of constant surveillance by American oil producers and Congressmen from states having a large number of independent oil producers."[33] This was a formidable obstacle to be circumvented.

On the other hand, when Stettinius, as acting secretary of state, discussed the matter with the navy,[34] he found the new secretary, James V. Forrestal, solidly behind him. To emphasize his position, Forrestal on December 11 wrote Stettinius a forceful letter, endorsing Stimson's letter of October 27. He argued that it was "distinctly in the strategic interest of the United States to encourage industry to promote the orderly development of petroleum reserves in the . . . Persian Gulf, thereby supplementing . . . Western Hemisphere sources and protecting against their early exhaustion." To this, he added a most interesting point. "The prestige and hence the influence of the United States is in part related to the wealth of the Government and its nationals in terms of oil resources, foreign as well as domestic . . . [and] . . . the bargaining power of the United States in international conferences . . . will depend in some degree upon the retention by the United

[33] Memorandum, "Some Observations on the Saudi Arabian Problem," by P. F. McGuire, November 3, 1944, 890F.51/11-344, RG59.

[34] Memorandum, Murray to Stettinius, November 11, 1944, and memorandum of meeting, Stettinius, Rayner, Undersecretary of War Patterson, and Undersecretary of the Navy Ralph Bard, November 13, 1944, file: "Saudi Arabia-Proposed Loan to," box 6, Records of the Petroleum Division, RG59; and memorandum of conversation, Stettinius, Murray, and Rayner, November 23, 1944, 890F.51/ 11-2344, RG59.

States of such . . . resources.'' For these reasons, the government
should develop a far-reaching program to increase worldwide use
of "oil produced in the Mesopotamian Basin" and to "protect
. . . [American] . . . holdings . . . in the Persian Gulf area.''[35]

The upshot of all of this was a combined State-War-Navy
recommendation to Roosevelt on December 22, outlining the
problem and requesting his approval for a reformulated program.[36]
To avoid the complications of a long-term airfield lease and an
overseas naval petroleum reserve, it was recommended that a
direct congressional appropriation be requested for interim aid to
the Saudi government. To back this up, and to replace it if Congress balked on a direct grant, the plan also called for an Export-
Import Bank loan and military airfield construction and training
missions, to demonstrate support for 'Abd al-'Aziz' efforts to
modernize his country.[37] Stettinius broached the plan to Roosevelt
in December, and obtained formal approval early in January
1945.[38] The Departments of State, War, and the Navy now had
presidential backing for a program of financial and technical aid
to Saudi Arabia. Only the details remained to be worked out.

The details required more than a year to resolve, partly because
every agency had a slightly different idea on how to proceed, and
partly because of rapid changes in personnel as the war came to

[35] Letter, Forrestal to Secretary of State, December 11, 1944, *FR(1944)*, V,
755-56. While all of this consultation was taking place, the new minister in Saudi
Arabia, William Eddy, continued to remind the department of 'Abd al-'Aziz'
unhappiness with the United States; telegrams, Eddy to Secretary of State, October
30 and November 28, 1944, *FR(1944)*, V, 751-53.

[36] Letters, Forrestal to Secretary of State, December 18, 1944, 890F.51/12-
1844, Stimson to Secretary of State, December 19, 1944, 890F.51/12-1944, and
memorandum, Murray to Secretary of State, December 20, 1944, 890F.51/12-
2044 RG59; and memorandum, Secretary of State to Roosevelt, December 22,
1944, *FR(1944)*, V, 757-58.

[37] Ibid. A further reason for dropping the idea of a naval petroleum reserve was
to disassociate the plan as far as possible from the ill will toward Ickes and his
Petroleum Reserve Corporation; memorandum of conversation, Stettinius et al.,
November 13, 1944, file: "Saudi Arabia-Proposed loan to," box 6, Records of
the Petroleum Division, RG59.

[38] Memorandum, Stettinius to Roosevelt, January 8, 1945, with marginal notation "ERS Jr., OK, FDR," *FR(1945)*, VIII, 847.

a close. What was in fact a long and intricate process can best be summarized by noting the final outcome of each of the recommendations. Consultations with key members of the House on March 8 and the Senate on May 17 convinced State that a direct congressional appropriation would be so fiercely opposed that it would not pass, and the idea was dropped.[39] The construction of an airfield met with more success. British objections were overcome by military and diplomatic overtures to London;[40] an agreement was negotiated with the Saudi government on August 6;[41] and a $4 million airfield was completed at Dhahran at American expense in early 1946 and ultimately turned over to the Saudi government.[42] And after much debate over how to ensure that the

[39] The March 8 meeting with House members was arranged by Acheson at the direction of the new State-War-Navy Coordinating Committee (SWNCC), which initially took charge of the Saudi project. Among others, the House leaders included the Speaker, the majority floor leader, and the chairman of the House Naval Affairs Committee, Carl Vinson; memorandum of conversation, Acheson et al., March 8, 1945, *FR(1945)*, VIII, 861-63; and memorandum, McGuire to Collado, March 24, 1945, 890F.51/3-2445, RG59. The May 17 Senate meeting included the majority leader, Alben W. Barkley, the chairman of the Committee on Finance, Walter F. George, and the chairman of the Senate Naval Affairs Committee, David I. Walsh; memorandum of conversation, Acheson et al., May 17, 1945, *FR(1945)*, VIII, 895-96. Accompanying Acheson at both meetings were Undersecretary of the Navy Bard, and Colonel George A. Brownell, executive officer to the assistant secretary of war for air.

[40] Memorandum, Dunn to Joint Chiefs of Staff, February 27, 1945, file: "091 Saudi Arabia-Miscellaneous—SWNCC," box 78, RG353; telegram, Stettinius to Eddy, April 17, 1945, memorandum, Department of State to British embassy, April 24, 1945, and telegram, Grew to Eddy, April 26, 1945, *FR(1945)*, VIII, 874-75, 881-82, 885.

[41] Memorandum, "Construction of a Military Airfield at Dhahran," SWNCC 19/13, June 14, 1945, file: "SWNCC 19-19/18, 091 Saudi Arabia I," box 8, RG353; memorandum, Grew to Truman, June 26, 1945, and letter, Eddy to Secretary of State, August 8, 1945, *FR(1945)*, V, 915-17, 943-50.

[42] Memorandum, Acheson to Truman, SWNCC 19/22, undated, memorandum, Truman to Acting Secretary of State, September 28, 1945, and telegrams, [Secretary of State James F.] Byrnes to Winant, November 19 and 30, 1945, *FR(1945)*, VIII, 957-58, 966-68, 971-72; letter, Brigadier General Donald P. Booth (special assistant to the undersecretary of war) to Senator James M. Mead (chairman, Special Committee Investigating the National Defense Program), December 11, 1945, file: "SWNCC 19/19-091 Saudi Arabia II," box 8, RG353; memorandum

Saudis would have sufficient dollar revenue to service a loan,[43] and how to cast the loan in terms that would not offend 'Abd al-'Aziz' religious objections to paying separate interest charges, an Export-Import Bank loan for $10 million was finally concluded on August 9, 1946.[44] The basic interest of the American government in the welfare of Saudi Arabia was well confirmed.

For our purposes, the most interesting part of this long and tortuous process was a side development that evolved into a matter of considerable importance. When Roosevelt approved the program of financial and technical aid to Saudi Arabia in January 1945, implementation was turned over to Dean Acheson and Will Clayton, who had just joined the department in December 1944 as assistant secretary of state for economic affairs. Clayton's views, therefore, began to have increasing impact. Thirty-six years as a highly successful southern cotton merchant had made him a firm supporter of the free enterprise system, free trade, and

of telephone conversation, Acheson with Halifax, December 20, 1945, telegram, Acheson to Eddy, December 20, 1945 (with note following), and telegram, Winant to Secretary of State, December 22, 1945, *FR(1945)*, VIII, 979-81, 983-87.

[43] Memorandum, Grew to Truman, May 23, 1945, memorandum of conversation, Truman et al., May 28, 1945, telegram, Grew to Eddy, June 27, 1945, telegram, Grew to Eddy, June 27, 1945, telegram, Byrnes to Winant, August 8, 1945, memorandum, Collado to Taylor (president of the Export-Import Bank), October 19, 1945, telegram, Acheson to Eddy, December 20, 1945, and telegram, Byrnes to Eddy, January 4, 1946, *FR(1945)*, VIII, 901-3, 917-18, 940-42, 960-63, 981-83, 999; memorandum, "Financial Aid to Saudi Arabia," attached to memorandum, R. L. Dennison to Secretary of the Navy, October 16, 1946, file: "Politico-Military Policy Division Subject Files (2)," 1946 and 1947, Records of the Office of the Chief of Naval Operations (hereafter cited as CNO Records).

[44] Telegram, Eddy to Secretary of State, March 21, 1946, telegram, Acheson to Sands (chargé in Saudi Arabia), May 3, 1946, telegram, Acheson to Childs (minister in Saudi Arabia), July 10, 1946, with n. 23, *FR(1946)*, VII, 740-41, 743-44, 746. With Truman's approval, Lend Lease for Saudi Arabia was continued into early 1946, beyond the general termination date for Lend Lease, in order to avoid a negative impact on 'Abd al-'Aziz; letter, Clayton to Crowley, September 4, 1945, and telegram, Acheson to Eddy, September 11, 1945, *FR(1945)*, VIII, 951-52; and memorandum, Byrnes to Truman, January 31, 1946, with notation "Approved 1-31-46, Harry Truman," file, "091-Saudi Arabia-Miscellaneous-SWNCC," box 78, Records of the State-War-Navy Coordinating Committee, RG59.

Cordell Hull. And four years of service in various wartime agencies before coming to State had done nothing to shake his conviction that the proper role of government was to create conditions favorable for business, but not to involve itself in the factional struggles of one group against another.[45] As already noted, when he took over the Bureau of Economic Affairs, he found the organization already deeply committed to postwar plans emphasizing free trade and multinational agreements designed to establish a healthy climate for business. Especially in tune with his thinking were the officers assigned to handle the details of the Saudi project—Emilio G. Collado, director of the Office for Financial and Development Policy,[46] and Paul F. McGuire, of the Division of Financial and Monetary Policy. Similar themes ran through most of their memoranda on the subject of Saudi finances, and, in fact, these memoranda mark the beginning of a new line of policy to replace that represented by the dying Anglo-American Agreement.

The Bureau of Economic Affairs had no quarrel with the basic objective of increasing Middle East production and keeping the Saudi concession in American hands. It was the means with which it took issue. As early as November 7, 1944, McGuire had opposed a direct governmental subsidy with the argument that "as long as the Arabian American Oil Company remains the sole beneficiary of the huge potential profits from Arabian oil, . . . that company can be expected under the American private enterprise system to provide risk capital to the extent of its financial

[45] Frederick J. Dobney, ed., *Selected Papers of Will Clayton*, pp. 1-17. Clayton was one of the founders and, for years, the prime moving force behind Anderson, Clayton, and Company. He came to Washington in 1940 at age sixty for a post in Jesse Jones' Reconstruction Finance Corporation and served in a series of agencies before joining State at the urging of Stettinius in 1944. He was assistant and then undersecretary of state for economic affairs until his retirement in 1948, and had considerable impact on postwar economic policy—including development of the Anglo-American Financial Agreement, the Marshall Plan, and the General Agreement on Tariffs and Trade (GATT).

[46] Collado left the Department of State in late 1947 for a position with Standard Oil (New Jersey) and later rose to a vice-presidency; telephone interview with Emilio G. Collado, from Cincinnati to New York, July 6, 1977.

capacity, appealing, if necessary, to the open capital market."
The department would be "treading on very dangerous ground
if it espouse[d] a doctrine that the 'national interest' justifie[d] use
of government funds to support operations of American private
companies abroad which compete with operations of other American private companies at home." An Export-Import Bank development loan was a different matter, however, because it would
be a straight commercial transaction, and "the fact that the oil
company might benefit indirectly from the stabilization and development of the Arabian economy would be only incidental, and
might therefore escape adverse comment."[47] On January 1, 1945,
Collado recommended to Clayton that, instead of a direct Congressional appropriation, State should press for an Export-Import Bank
loan, additional advances by Aramco to make up the Saudi deficit,
and for continuation of Lend Lease through 1945. As a postscript,
he suggested talking with Aramco about "*bringing in additional
companies as the one company may not have sufficient markets
to handle the concession* (emphasis added)."[48]

On February 10, McGuire firmly endorsed this idea. In a memorandum to Collado, he argued that, "If the Arabian American
Oil Company actually does not have enough financial resources
to permit it to function as an adequate vehicle for the national
interest now entrusted to it, perhaps it should be required to permit
other American oil companies to participate in the concession.
This would not only supplement the financial resources available
for the development of Arabian oil, but would provide additional
marketing facilities, give a greater portion of the domestic industry
an interest in the development of this oil, and . . . reduce the
potential opposition to an integrated use of the world's petroleum
reserves in the interest of security and conservation."[49] And on
April 7, Clayton speculated that there might be "compelling reasons of which I am not aware for believing that the oil concession
cannot be *adequately protected by the private interests entitled to*

[47] Memorandum, "Conclusions on the Saudi Arabian Problem," by P. F.
McGuire, November 7, 1944, 890F.51/11-744, RG59.

[48] Memorandum, Collado to Clayton, January 1, 1945, 890F.51/1-145, RG59.

[49] Memorandum, McGuire to Collado, February 10, 1945, 890F.51/2-1045,
RG59.

the profits from its exploitation. But if so, doubt is cast upon the adequacy of the American private enterprise system in the international field."[50] The Bureau of Economic Affairs clearly believed that the matter should be handled by private capital, and that additional companies should be brought in to handle the task if Aramco lacked adequate resources to do the job. But up to this point there had been no discussion of the subject with Aramco or any of the other companies involved.

By May, however, discussion of the possibility of bringing in Jersey and Socony had spread at least as far as the Navy Department, Socal, Aramco, and the British Foreign Office. It is impossible from the accessible record to determine whether Socal and Aramco first broached the subject to the Departments of State and the Navy and asked for their comments, or vice versa, but in the early summer of 1945 State and Navy representatives solidly endorsed the idea of a merger in informal contacts with Socal and Aramco.[51] The British Foreign Office received a report quoting Loy Henderson of NEA to the effect that "the Standard Oil Company of California . . . had become rather staggered at the magnitude of the proposition it had to handle . . . [and] . . . would be quite relieved to bring in other oil companies to share with them . . . [but were blocked by] . . . the Red Line Agreement . . . [which was now] . . . out of date and even prejudicial."[52] Undersecretary of the Navy Ralph Bard discussed the subject with

[50] Clayton to [Assistant Secretary of State James] Dunn, April 7, 1945, *FR(1945)*, VIII, 869-71.

[51] Follis interview. Socal officers were in Washington frequently during the later part of the war renegotiating the Elk Hills Naval Petroleum Reserve contract with Undersecretary Bard and his special assistant, Carl McGowan; telephone interview with Judge McGowan, Cincinnati to Washington, D.C., July 6, 1978. Follis recalled numerous queries as to whether Socal was doing everything possible to keep the concession in American hands. When he pointed out that the company had limited markets and could have antitrust problems if it got together with other companies to increase sales, he recalled that State and Navy personnel said that it would be an "outrage" if antitrust considerations endangered the concession, and promised complete support in discussing the issue with the Department of Justice.

[52] Foreign Office Minutes by R. M. Butler, July 31, 1945, reporting a conversation with Henderson "the first week in July," V50385, W10686/12/76, FO/371.

James Terry Duce in May, and Duce told him that "Arabian American would welcome an approach to . . . [Jersey and Socony-Vacuum] . . . by the Government with a view to seeing what could . . . be worked out."[53] Secretary of the Navy Forrestal expressed the opinion that he didn't "care which American company or companies develop[ed] the Arabian reserves" so long as they were "*American*."[54] Finally, on June 22, in a meeting with Director of the Office of War Mobilization and Reconversion Fred M. Vinson on the subject of aid to Saudi Arabia, Bard agreed that he "would discuss with other American oil companies the question of securing their cooperation in increasing the markets for Arabian oil."[55] Not only was the government in favor of broadening participation in the concession, but it was on the verge of serving as a marriage broker.

At this point, there is a tantalizing gap in the evidence. Shortly after the meeting just commented upon, Bard left government service and Vinson was appointed secretary of the treasury.[56] Clayton and Collado became involved in the Potsdam Conference, and a number of other participants left government service or changed jobs.[57] None of the surviving documents, and none of the participants who could be located recall that an actual approach was made by the government to Jersey or Socony-Vacuum.[58]

[53] Memorandum, McGowan to Bard, May 31, 1945, file: "36-1-30," Forrestal Office files, Record Group 80 (hereafter cited as RG80).

[54] Memorandum, Forrestal to Secretary of State, August 1, 1945, file: "36-1-30," Forrestal files, RG80. The memorandum was prepared, but apparently not sent.

[55] Memorandum of conversation, Vinson, Bard, John J. McCloy (assistant secretary of war), Edward F. Prichard (special assistant to Vinson), Edward S. Mason (special assistant to Clayton), Carl McGowan (special assistant to Bard) et al., June 22, 1945, 890F.51/6-2245, RG59.

[56] Bard retired at age sixty on June 30, 1945; Biographical sketch, Ralph A. Bard, Naval History Division, Navy Yard, Washington, D.C. Vinson was confirmed as secretary of the treasury on July 16, 1945; "Introduction to the Guide of the Fred M. Vinson Collection," Fred M. Vinson Papers.

[57] Dobney, p. 10; Collado interview; letter to the author from Edward S. Mason, August 19, 1977; and letter to the author from Edward F. Prichard, Jr., May 22, 1978.

[58] Collado interview; McGowan interview; Mason letter; telephone interview with Robert R. Nathan, former deputy director of the Office of War Mobilization

Jersey management may have indirectly become aware of the government's position,[59] but there is no evidence that it was ever formally notified.

Whether or not government endorsement was conveyed to Jersey as well as to Socal is not critical to the point being developed, however. The fact is that a major shift in government policy occurred in the early part of 1945.[60] The objectives of increasing Middle East production to conserve Western Hemisphere reserves

and Reconversion, in Washington, D C., July 2, 1979; Prichard letter, and letter to the author from Willard L. Thorp (who took over Mason's responsibilities at State when Mason left government service July 1, 1945), September 23, 1977. The possibility of approaching Jersey and Socony-Vacuum for some type of help in solving Aramco's marketing problem was kept alive in intragovernment discussions at least until the first of August, 1945, but there is no indication that a contact had been made up to that point, and no indication beyond that point as to the final disposition of the idea; memorandum, McGowan to Acheson, July 13, 1945, 890F.6363/7-1345, RG59; letter, Prichard to Vinson, July 14, 1945, and memorandum, Clark M. Clifford (at the White House) to Captain Vardaman, July 18, 1945, file: "Report on Petroleum Policy-Saudi Arabia," box 176D, item 16, Record Group 250, Records of the Office of War Mobilization and Reconversion, (hereafter cited as RG250); memorandum, Forrestal to Secretary of State, August 1, 1945, file: "36-1-30," Forrestal Office files, RG80; memorandum, Assistant Secretary of the Navy for Air John L. Sullivan to Forrestal, "Saudi Arabia-Status" undated, but text places it sometime after mid-July, 1945, file: "Saudi Arabia," box 89, James V. Forrestal Papers (hereafter cited as Forrestal MSS); memorandum, "Comment Annexed to Navy Department's Memorandum of Concurrence in SWNCC165," by A. L. Gates, August 8, 1945, file: "QB(478)/A3-1, JJ7-11," RG80; memorandum, McGowan to Carter et al., August 25, 1945, file: "36-1-30," Forrestal Office files, RG80; and memoranda by Harold Stein (Office of War Mobilization and Reconversion), September 18 and 19, 1945, file: "Report on Petroleum Policy—Saudi Arabia," box 176D, item 16, RG250.

[59] Jersey management was discussing broadening its supply position in the Middle East in mid-1945, but the minutes of its executive committee do not indicate any knowledge of government endorsement at this point; index to records of executive committee, entry for "Arabian American Oil Company, 1945," excerpts from records of the Exxon Corporation provided to the author by Professor Bennett H. Wall, director, Center for Business History Studies, Tulane University, New Orleans, Louisiana (hereafter referred to as Exxon Records). Professor Wall is preparing a history of Exxon since 1950, to follow Larson, Knowlton, and Popple's volume on the *History of Standard Oil Company (New Jersey)*. On behalf of the author, he graciously searched the few surviving Exxon records from the 1940s for all references to the Aramco acquisition.

[60] It should be clear that the term "policy" is used in the de facto sense, since

and of keeping the Saudi concession solidly in American hands by assuring an adequate return for 'Abd al-'Aziz remained as firm as ever. But the means changed. Partly because of the Texas independents' opposition to international control of production, and partly because of the growing influence of Will Clayton, the Department of State discarded the Anglo-American Petroleum Agreement as an instrument of policy, and turned to support of private initiative to achieve those objectives. Endorsement of broader participation in the marketing of Saudi oil was the first step in this direction.

*

CONTRARY to conventional wisdom and to the findings of the Church Committee in 1975,[61] it was Socal rather than Jersey that made the first move to broaden ownership of Aramco. Jersey had been interested in some type of marketing arrangement for a long time, and Jersey had helped create the conditions that made the move desirable to Socal and Texas. But the actual suggestion that a partnership be formed was made by R. Gwin Follis (later president and board chairman) of Socal at the direction of H. C. Collier to Jersey Vice-President Orville Harden in early 1946.[62] Much spec-

---

there was no formal, high-level review wherein the shift was decreed. The change was evolutionary and a product of changing circumstances.

[61] U.S. Congress, Senate, Committee on Foreign Relations, Subcommittee on Multinational Corporations, *Multinational Oil Corporations and United States Foreign Policy, Report*, p. 46. The Subcommittee chairman was Senator Frank Church of Idaho. The 1952 FTC report, *The International Petroleum Cartel*, also stated that negotiations "were initiated by Jersey and Socony-Vacuum" (p. 120).

[62] Follis interview. Follis did not recall the exact date of this contact, but a contemporary Socal memorandum notes that informal conversations with Jersey took place in "the latter part of May, 1945" (memorandum, Follis to Stoner et al., August 18, 1946, U.S. Congress, Senate, Committee on Foreign Relations, Subcommittee on Multinational Corporations, *Multinational Corporations and United States Foreign Policy, Hearings*, pt. 8, p. 96); the minutes of Jersey's executive committee for May 28, 1976, note that "Holman and Harden informed the Committee of discussions they had had with representatives of other companies concerning . . . obtaining a participation in the subsidiaries of such companies . . . which . . . hold important petroleum concessions in the Middle East" (index to records of executive committee, entry for "Arabian American Oil Company,

ulative material has been written about the reasons why Socal and
Texas were willing to let Jersey and Socony in, but contemporary
documents and interviews with corporate officers close to the
transaction clearly indicate that the principal reason was business
conservatism on the part of both Collier and Rodgers. There was
more oil in Saudi Arabia than either company could foresee marketing during the life of the concession; $80 million in risk capital
had already been advanced and the pipleine would require another
$100 million; a fight with the other majors to radically increase
their share of the market would be both expensive and risky; and
the concession was seen as not really secure unless markets were
increased rapidly enough to satisfy 'Abd al-'Aziz' desire for a
larger income.[63] The companies had lost their bid for government
financing of the refinery and the pipeline, and the Departments

1946,'' Exxon Records); and there is no mention of the subject in the minutes of
the Socal executive committee or board of directors meetings prior to May, 1946,
(letter to the author from W. K. Morris, vice-president, Standard Oil Company
of California, August 15, 1978). Nothing in the accessible record contradicts
Follis' memory that he made the initial contact, and his memory tends to be
confirmed by a legal brief, prepared with considerable attention to accuracy by
Aramco attorney George Ray in 1956, which states that ''the owners of Aramco
invited Standard Oil Company of New Jersey . . . and Socony Mobil Oil Company
. . . to acquire partial ownership of Aramco'' (Saudi Arabia, *Arbitration Between
the Government of Saudi Arabia and Arabian American Oil Company*, V, 13; and
letters to the author by George W. Ray, Jr., February 27 and March 20, 1978).
Apparently, the initial contact was made informally sometime prior to May 28
and then brought formally to the respective executive committees for authority
to proceed.

[63] Follis interview; Long interview; memorandum, Follis to Stoner et al., August
18, 1946, and confidential memorandum for Texas board meeting of August 16,
1946, U.S. Congress, Senate, Committee on Foreign Relations, Subcommittee
on Multinational Corporations, *Multinational Corporations and United States
Foreign Policy, Hearings*, pt. 8, pp. 96-111; Theodore L. Lenzen, ''Inside International Oil'' (unpublished typescript manuscript based in part on contemporary
notes on his participation in these negotiations as an aide to Collier), pp. 36-38,
and interview with Lenzen in Atherton, California, August 24, 1977. Socal placed
more emphasis on the need for markets and Texas placed more emphasis on the
need for funds in their planning, but these were simply two dimensions of the
same problem; letters to the author by W. K. Morris, vice-president, Standard Oil
Company of California, March 15, 1978, and A. C. De Crane, Jr., senior vice-
president, Texaco, Inc., March 29, 1978.

of State and the Navy were now urging expanded participation in order to protect the concession. Under these circumstances, the prudent thing to do was to bring in Jersey and Socony, let them fund the pipeline, and let them sell enough of the "surplus" oil to satisfy 'Abd al-'Aziz' thirst for royalties. Such a move would both decrease and spread the risk without damage to the long-term profit picture for either Socal or Texas. There was to be internal opposition toward this line of reasoning within Socal, and some initial difficulty with Rodgers over the terms of the deal, but Collier carried the day and the offer was made.

As already noted, Jersey and Socony had been interested in some type of participation in Saudi oil since the early 1930s but had been effectively blocked by the Red Line Agreement and the attitudes of Gulbenkian and the Compagnie Française des Pétroles (C.F.P.). Jersey had been crude-short ever since dissolution of the original Standard Oil Trust in 1911, and studies of its projected postwar position undertaken in the latter part of the World War II indicated a growing need for Eastern Hemisphere sources to supply Eastern Hemisphere markets.[64] By June of 1945, Jersey's executive committee had begun to discuss the need for "acquiring a larger participation in the petroleum reserves in the Middle East," and by February of 1946, it had begun consideration of "various types of arrangements . . . to acquire an increased participation in such supplies."[65] Aramco's basic position was well known throughout the industry, and when Follis approached Harden on a possible combining of interests, the suggestion fell on fertile ground.

There is another factor that is more difficult to evaluate. This is the question of whether or not the original merger, in and of

[64] Interview with former Jersey director Howard W. Page (who worked on these studies) in New York, May 2, 1977; index to records of the executive committee, entry for "Arabian American Oil Company, 1945," Exxon Records; *Standard Oil Company (New Jersey) and Middle Eastern Oil Production*, March, 1945 (advance proof), file 36-1-30, Forrestal Office files, RG80; and Larson et al., p. 734. A summary of Jersey's worldwide production and sales during this period is given in Appendix C.

[65] Index to records of the executive committee, entries for "Arabian American Oil Company," 1945 and 1946, Exxon Records.

itself, constituted a violation of American antitrust laws. It was obvious to all parties that if the markets of Jersey and Socony were denied to Aramco, the pressure from 'Abd al-'Aziz to increase production might force Socal and Texas into a fight for a larger share of the market, and with the low cost of Saudi crude, this could seriously destabilize world markets.[66] This point was discussed at least once within both Jersey and Texas in 1946,[67] but the documentary record indicates that this was a secondary consideration, if one at all. To further complicate matters, the companies took the position that they were covered by the Webb-Pomerene Act of 1918, which made combinations of American exporters overseas perfectly legal, provided this did not produce a restraint of trade *within the United States*.[68] Even if stabilization of world markets had been the primary reason for buying into Aramco, there would remain the question of whether or not it could be demonstrated that the merger in and of itself affected trade within the United States. Subsequent critics in the Federal Trade Commission and the Anti-Trust Division of the Department of Justice charged that it could,[69] and this was one of a long list of allegations that led to the antitrust suits initiated against Jersey, Socony, Socal, Texas, and Gulf in 1952. As will be seen later,

[66] U.S. Congress, Senate, Select Committee on Small Business, Subcommittee on Monopoly, *The International Petroleum Cartel*, pp. 120-21.

[67] Testimony of Barbara J. Svedberg, Antitrust Division, Department of Justice, U.S. Congress, *Multinational Corporations*, pt 7, p. 82; and confidential memorandum for Texas board meeting of August 16, 1946, U.S. Congress, *Multinational Corporations*, pt. 8, p. 111.

[68] *An Act to Promote Export Trade and For Other Purposes, Statues at Large*, vol. 60, chap. 50 (1918), pp. 516-18.

[69] U.S. Congress, *International Petroleum Cartel*, pp. 119-22; and Svedberg testimony, U.S. Congress, *Multinational Corporations*, pt. 7, pp. 60-63 and 81-83. These charges are vehemently summarized in John Blair, *The Control of Oil*, pp. 34-42. Blair spent much of his life with the FTC and Senate investigating committees crusading against concentration in the oil industry, and if anyone could prove a point, it would be Blair. His work, however, requires careful scrutiny. Although making a number of valid points, he frequently goes beyond the actual evidence by citing charges made in previous studies (including his own) as though they were fact, and by extensive use of logical inference rather than documentary evidence.

those cases were sharply limited by Presidents Harry Truman and
Dwight Eisenhower for national security reasons during the Iranian crisis of the 1950s and ultimately settled by consent decrees
or dropped.[70] As a result, the charges have never been settled in
a court of law. From the point of view of the historian, however,
it must be reported that the accessible corporate documentary
records support the assertions of corporate officers that the principal reason for Jersey's interest in 1946 was a need for additional
sources of supply. This, of course, does not clear the companies
of charges arising from actions taken by the Aramco board *after*
the merger, but those will be dealt with later.

Of more legal significance in 1946 was an interpretation of
British law that played a significant role in the negotiations that
followed. When Hitler overran France in 1940, the holdings of
Gulbenkian and C.F.P. in the Iraq Petroleum Company had been
sequestered by the British custodian of enemy property because
both were domiciled in territory controlled by the enemy.[71] In the
process of rearranging its legal affairs to deal with this situation,
the IPC board in London received an unexpected opinion from
Barrister A. Andrewes Uthwatt that the entire 1928 Group (Red
Line) Agreement might have been "frustrated" (rendered null
and void) by the "supervening illegality" of becoming a contract
with enemy aliens, which was illegal at that point.[72] Since the
other terms of the agreement called for relinquishment of rights

[70] There is an excellent analysis of these cases in Burton I. Kaufman, *The Oil
Cartel Case: A Documentary Study of Antitrust Activity in the Cold War Era.*

[71] Letters, Harold Christie (Custodian's Office, London) to Secretary, Iraq Petroleum Company, July 13 and September 20, 1940, file 32907, FTC Records.
Gulbenkian's designation as an "enemy alien" ended in 1943 when he moved
to Lisbon; C.F.P.'s designation continued until the end of the war in Europe.

[72] Letters, C. Stuart Morgan (Near East Development Company, New York)
to Wallace Murray, September 12, 1940, 890G.6363-T-84/652, and March 3,
1941, 890G.6363-T-84/653, with enclosures, RG59. The term "supervening illegality" is not in the Uthwatt opinion itself; it was an embellishment added in
later discussions of the issue; interview with former Jersey director David Shepard
in New York, May 3, 1977. Uthwatt happened to be an authority on the doctrine
of "frustration," and he apparently added this opinion gratuitously to his reply
to the other questions asked.

to oil that participants could not lift (as was the case with Gulbenkian and C.F.P. during the war), and since this served the interests of the other companies at the time, they elected not to act on the Uthwatt opinion and to let the matter ride.[73] At the war's end, however, when Gulbenkian and C.F.P. attempted to reassert rights to the quantities of oil they would have received during the war, the whole question of whether or not the Red Line Agreement had been ''frustrated'' was resurrected in British legal circles.[74] Sometime in late 1945 or early 1946 Jersey General Counsel Edward F. Johnson called this issue to the attention of Orville Harden as a way in which Jersey might extricate itself from the agreement, if it so desired. Harden expressed little interest at the time, but sometime later (presumably after the overture from Follis) he came back and asked Johnson to pursue the matter in depth. Johnson promptly left for England, where he obtained three separate legal opinions that the agreement had indeed been ''frustrated.''[75] This became the legal basis for Jersey and Socony's dealings with Gulbenkian and C.F.P. for the next two years.

While Harden and Holman were clearing the legal decks for negotiations, Collier had to contend with a revolt within his own ranks. Almost to a man, Aramco field management opposed the

[73] Letter, R. W. Sellers (Socony, London) to H. F. Sheets (Socony, New York), January 21, 1941, and telegram, Near East Development to Piesse (London), February 12, 1941, 890G.6363-T-84/653, RG59.

[74] Memorandum of conversation, Sheets, Hardon, Clayton, Rayner et al., March 22, 1945, 890G.6363/3-2245, RG59; letter, H S. Gregory (Trading with the Enemy Department, Treasury and Board of Trade) to H.W.A. Freese-Pennefather (Foreign Office), April 13, 1945, V50381, W5211/12/76, FO/371; Foreign Office minute, June 7, 1945, V50382, W7519/12/76, FO/371; and minutes of Jersey Executive Committee meeting, August 1, 1946, file 33369, FTC Records.

[75] Interview with Edward F. Johnson, in Scarsdale, New York, July 12, 1977, and letters to the author from Johnson, August 2, and October 1, 1977. Johnson recalled informing Harden of the doctrine of ''frustration'' but did not recall knowing of the earlier Uthwatt opinion. However, since he made several trips to England during this period, it is highly likely that this was where he first became aware of the doctrine. The three opinions were obtained by Jersey Soliciter Montagu Piesse from Barristers Gerald Gardiner, D. N. Pritt, and Sir Ballentine Holmes; Johnson interview and FR(1946), VII, 32.

idea of a sale.[76] They were production men responsible for one of the greatest discoveries in history, and they wanted to go it alone. McPherson even spoke of Aramco building its own marketing organization and selling for its own account rather than through Caltex or the parents.[77] The leader of the opposition, however, was Reginald Stoner in San Francisco. On June 10 he wrote a long, forceful memorandum to Collier and Follis opposing the sale. Arabian fields, he said, were "the most prolific" in the world, with "the cheapest" production costs anywhere. Caltex was already considering acquiring Texas outlets in Europe in order to operate there directly, and Jersey was so tied in with Shell and others in worldwide production arrangements that Aramco was in a much better position to expand rapidly alone, "not because we already have the markets, but because we have . . . cheap oil available right now." And if Caltex couldn't do it alone, it was "common knowledge that all large companies in the United States" were "seeking foreign oil," including "Phillips, Barnsdale, Atlantic Refining, Standard of Indiana, and Sinclair."[78] Collier, however, was the chief executive officer of Socal, and a strong personality. He had a study done for the board of directors that "proved" that, over the life of the concession, Socal would net a larger profit with, than without, Jersey,[79] and he simply overrode Stoner's objections. The profitability study was a highly speculative one and not the real issue. The real issue was how the business was going to be run, and Collier had already decided on the more conservative course.[80]

The next problem was with Rodgers. He had been designated

---

[76] Ohliger interview.   [77] Barger interview.

[78] Memorandum, Stoner to Collier and Follis, June 10, 1946, U.S. Congress, *Multinational Corporations*, pt. 8, pp. 84-89.

[79] Memorandum, Lenzen et al., to Follis, June 18, 1946, U.S. Congress, *Multinational Corporations*, pt. 8, pp. 89-93.

[80] Follis interview; Long interview. Blair speculates that Stoner was overridden because Rockefeller interests wanted the deal consummated, and that "financial interest group was by far the leading stockholder in both Socal and Exxon" (p. 39), but he produces no evidence to substantiate the speculation. The explanation offered here is consistent with the memory of all of those interviewed and with all of the documents that could be located.

to handle the preliminary negotiations with Harden, and—true to form—he opened by placing a $650 million price on the one-third interest Socal and Texas proposed to sell. This was considerably more than Harden had been led to expect, and he bypassed Rodgers and complained directly to Socal. As a result, Follis went to New York and convinced Rodgers to bring the asking price down to $250 million. Further discussions led to a concrete offer to Jersey on July 29 and 30, in meetings attended by Collier, Follis, Stoner, Rodgers, Klein, Holman, Harden, and a number of others from all three companies.[81] Negotiations now began in earnest.

In August, Jersey decided to formally notify its partners in IPC that it considered the Red Line Agreement null and void and to offer to negotiate a new working agreement free of the old restrictive clauses.[82] This action required the cooperation of Jersey's partner in the Near East Development Company—Socony-Vacuum, and on August 27, Harden and board chairman Harold F. Sheets of Socony went to Washington to ensure State Department support for their position. They met with Clayton, Rayner, and George McGhee and were assured that present government policy opposed restrictive agreements of any kind, and that support would be forthcoming if required. The negotiations with Socal and Texas were not mentioned at this meeting.[83] Harden and Sheets then went to London to inform the IPC board of the position they were taking. All of the other groups stated formal disagreement, but Shell told Harden privately that it would cooperate in the drafting of a new agreement.[84] That left Anglo-Iranian, C.F.P. and Gul-

[81] Memorandum, Follis to Stoner et al., August 18, 1946, U.S. Congress, *Multinational Corporations*, pt. 8, pp. 96-100.

[82] Minutes of Jersey executive committee meeting, August 21, 1946, file 33369, FTC Records.

[83] Memorandum of conversation, Hardin [sic], Sheets, Clayton, McGhee, and Rayner, August 27, 1946, *FR(1946)*, VII, 31-34; memorandum (by Sheets), August 28, 1946, U.S. Congress, *Multinational Corporations*, pt. 8, p. 112; and letter, Secretary of State to Chargé in the United Kingdom, November 29, 1946, *FR(1946)*, VII, 38-40.

[84] Minutes of Jersey executive committee meetings of October 9 and December 9, 1946, file 33369, FTC Records; letters, R. W. Sellers to Sheets, September

benkian's representative in real disagreement, in that order of
intensity.

In parallel with all of this, Jersey had been conducting side
discussions with Socony-Vacuum on the possibility of Socony
also coming into Aramco, and this subject was broached to Socal
and Texas in late October. It met with approval, and Socony
entered the negotiations on November 6.[85] Socony's board of
directors, however, was of mixed mind. Sheets, a marketing man,
favored taking up a share equal to that of Jersey, but Socony
president Brewster Jennings was more cautious. Along with John
C. Case, who handled a portion of the negotiations, he feared that
Middle Eastern oil "was not absolutely safe" and the company
"ought to put more money in Venezuela."[86] The resolution was
an agreement that Socony would seek a smaller interest, 10 percent
compared with 30 percent for Jersey, which was about equal to
the percentage of Aramco's production that each company thought
it could market.[87]

In November, the Jersey executive committee decided that the
time had come to officially inform the Departments of State, the
Navy, the Interior and Justice of the project.[88] Rayner had been
briefed confidentially in July,[89] but this round of meetings was

17, 1946, and Sellers to Jennings, October 2, 1946, U.S. Congress, *Multinational
Corporations*, pt. 8, pp. 112-16.

[85] Letters from Sheets to Laurence B. Levi, July 20, 1945, C. V. Holton to
Jennings, October 28, 1945, and G. W. Orton to Rodgers et al., November 8,
1946, U.S Congress, *Multinational Corporations*, pt. 8, pp. 94-95 and 116-18;
Lenzen, "Inside International Oil," p. 40

[86] Interview with John C Case, in Keene Valley, New York, July 11, 1977.
Socony counsel C. V. Holton considered the proposal technically legal, but had
reservations on the grounds that he could not "believe that a comparatively few
companies for any great length of time . . . [were] . . going to be permitted to
control world oil resources without some sort of regulation"; letter, Holton to
Jennings, October 28, 1946, U.S. Congress, *Multinational Corporations*, pt. 8,
116-17.

[87] Letter, Orton to Rodgers et al., November 8, 1946, U.S. Congress, *Multinational Corporations*, pt. 8, pp. 117-18.

[88] Minutes of Jersey executive committee meetings of November 16 and 19,
1946, Exxon Records.

[89] Letter, Sheets to Laurence B. Levi, July 20, 1946, U.S. Congress, *Multinational Corporations*, pt. 8, pp. 94-95.

intended to ensure support and to protect the companies against unexpected and uninformed criticism. In late November, Holman and officers from the other companies met with Secretary of the Navy Forrestal, Assistant Secretary of State Donald S. Russell (substituting for Clayton), and Ralph Davies, now acting head of the Oil and Gas Division of the Department of the Interior, and received their full endorsement.[90] On December 3, Harden followed up with a detailed discussion with Clayton, McGhee, and Henderson.[91] On December 4, General Counsels Edward Johnson of Jersey, George V. Holton of Socony, and Harry T. Klein of Texas met with Attorney General Tom Clark, who offered no objections to the plan.[92] (Johnson, Holton, and Klein discussed the proposed merger with Clark again on January 16 and again received no objections.)[93] On December 6, Holman, Brewster, and Rodgers met with Secretary of War Robert Patterson and Secretary of the Interior Julius Krug and found that "their reaction was favorable" also.[94] After all, the merger was completely consistent with the position that had been adopted by State and the Navy a year earlier. It was a private initiative that would increase the markets for Saudi oil, decrease the drain on Western Hemisphere reserves, and keep the concession firmly in American hands.

The deal was close to consummation, and Jersey took one more step to strengthen its position. On December 20, it signed and announced a twenty-year contract with Anglo-Iranian for a sizable

[90] Minutes of Jersey executive committee meeting of November 21, 1946, Exxon Records.

[91] Memorandum of conversation, Harden, Clayton, McGhee, and Henderson et al., December 3, 1946, *FR(1946)*, VII, 40-43.

[92] Minutes of Jersey executive committee meeting of December 5, 1946, Exxon Records.

[93] Minutes of Jersey executive committee meeting of February 6, 1947, Exxon Records. The Antitrust Division of the Department of Justice did not begin to analyze the merger documents until a year after they had been delivered to the attorney general, and then only after being prodded to do so by the Senate Special Committee Investigating the National Defense Program; U.S. Congress, Senate, Committee on Foreign Relations, Subcommittee on Multinational Corporations, *Multinational Oil Corporations and U.S. Foreign Policy, Report*, pp. 49-50.

[94] Minutes of Jersey executive committee meeting of December 9, 1946, Exxon Records.

quantity of either Iranian or Kuwaiti oil (at the discretion of Anglo-Iranian), and an agreement to form a jointly owned Middle East Pipeline Company and construct another pipeline from Abadan across Iraq and Syria to the Mediterranean to move oil to the European market.[95] The agreement had three purposes: it helped to alleviate Jersey's long-term supply problem; it bought Anglo-Iranian's acquiescence to modification of the Red Line Agreement; and it served as a bargaining lever with Socal and Texas—demonstrating that Jersey could obtain Middle Eastern oil from sources other than Aramco.[96] The supply contract was a straight business transaction, but it is difficult to tell how serious Jersey really was about the "Middle Eastern Pipeline." That project was dropped in 1950 on the grounds that it was too difficult to obtain permission from Iraq to transport Iranian oil across Iraqi territory.[97] The chief effect of the agreement in late 1946, however, was to remove Anglo-Iranian as an obstacle to the Aramco merger—only C.F.P. and Gulbenkian were now left.

The French had been thinking over their position for some time, and announcement of the Anglo-Iranian supply contract appears to have tipped the scales. On January 4, 1947, the French ambassador in the United States, Henri Bonnet, handed Will Clayton,

[95] "Heads of Terms Made Between Standard Oil Co. (New Jersey) ('Buyers') and Anglo-Iranian Oil Co. Ltd. ('Suppliers'), December 20, 1946," U.S. Congress, *Multinational Corporations*, pt. 8, pp. 119-21. The supply contract was for an estimated 133 million tons over the twenty-year period.

[96] Minutes of Jersey executive committee meetings of August 29, October 24 and 25, November 8, 15, 19, and December 23, 1946, FTC Records; memorandum of conversation, Harden, Clayton et al., December 3, 1946, and telegram, Acting Secretary of State to Chargé in the United Kingdom, December 4, 1946, *FR(1946)*, VII, 40-44; Page interview.

[97] Telegram, Moose (in Iraq) to Secretary of State, January 13, 1947; memorandum of conversation, Harden, McGhee, Rayner et al., February 3, 1947, aide-mémoire, British embassy to Department of State, July 9, 1947, *FR(1947)*, V, 633, 639-42 and 660-62; memorandum, Henderson to Secretary of State, February 10, 1948, letter, Secretary of State George C. Marshall to George Koegler (Jersey counsel), March 8, 1948, and editorial note, *FR(1948)*, V, 5-7 and 31; U.S. Congress, *International Petroleum Cartel*, pp. 154-60; and Shepard interview. Before the project was cancelled it had also been agreed that Socony would join in this project as a junior partner to Jersey.

now undersecretary of state for economic affairs, a formal protest over the American companies' "unilateral denunciation of the [Red Line] Agreements of 1928.''[98] The French position was clear and quite understandable. In part, Gallic pride had been wounded by first learning of the Aramco negotiations from press reports in December, but there was a practical consideration, too. The Iraq Petroleum Company was the only source of supply for the Compagnie Française des Pétroles, and by the 1928 agreements, each partner in IPC took a fixed percentage of the annual output, at cost. C.F.P. feared that Jersey, Socony and Anglo-Iranian might work to increase production in Saudi Arabia and Iran, and, as partners in IPC, retard it in Iraq. To avoid being placed at this competitive disadvantage, C.F.P. took the position that the Red Line Agreement was still valid and demanded the right to partic ipate in Aramco along with Jersey and Socony. In defense of its position, it initiated legal action in February in the British courts to secure an injunction and to reaffirm the validity of the agreement.[99] The French government, which held a major interest in the company, backed this up with a strong diplomatic protest to the United States. Gulbenkian was in a similar position and threatened parallel action in the British courts. He, however, simply wanted to drive the best bargain possible for himself.[100] When consulted on these developments, 'Abd al-'Aziz made it clear that he would not agree to sale of a partial interest in Aramco to any company other than an American one,[101] and this created a situ-

[98] Bonnet to Secretary of State, January 4, 1947, *FR(1947)*, V, 627-29. The note of January 4 was backed up by a more detailed one delivered on January 13, 1947 (890.6363/1-447, RG59).

[99] Minutes of Jersey executive committee meetings of December 26, 1946, January 9, and March 5, 1947, Exxon Records; memorandum of (IPC) group meeting in London, December 19, 1946, letter, Gulbenkian to Near East Development Corporation, January 6, 1947, and letter, Denton et al., to Piesse, January 7 and 8, 1947, U.S. Congress, *Multinational Corporations*, pt. 8, pp. 123 and 126-28; memorandum of conversation, Bonnet and Clayton et al., January 10, 1947, *FR(1947)*, V, 632-33; letter, Caffery (Paris) to Secretary of State, January 13, 1947, 891.6363 AIOC/1-1347, RG59; Shepard interview.

[100] Case interview.

[101] Memorandum, Henderson to Acheson, January 16, 1947, *FR(1947)*, V, 634-35.

ation where the deal could not be consummated until the legal issue was completely settled—in court or out.

The State Department now had to respond to the French government, and it undertook a full-scale review to decide on the exact position that it should take. A staff paper by John Loftus pointed out that the Aramco transaction, the Jersey-Anglo-Iranian contract, and a concurrent long-term Gulf-Shell contract to supply Shell with Kuwaiti oil all served "United Sates national interest by hastening the development of Middle Eastern oil" and reducing the "drain upon Western Hemisphere oil resources." In addition, "the ARAMCO and the Gulf-Shell deals . . . [would] . . . facilitate development of extensive American holdings in the Middle East." On the other hand, the department "would be subject to legitimate criticism" if it failed to recognize "that the arrangements with their interlocking connections and combining of interests" might reduce competition and facilitate "market and price policies incompatible with general United Sates trade objectives." Since "neither blanket approval or blanket disapproval . . . [was] . . . feasible," the department should resort to "the 'no objection' formula" in discussing the matter with the oil companies.[102] The influences of Clayton and wariness about criticism from the independents were both evident in Loftus' argument. This line of reasoning led Paul Nitze, deputy director of the Office of International Trade Policy, to suggest to Clayton that the entire matter might be resolved to everyone's satisfaction by Jersey selling its interest in IPC to Socony and going into Aramco alone.[103] The idea was well enough received within the department for Nitze to broach it to Harden and Jennings during a meeting on March 7 called to discuss the French note. Harden told him that the idea had already been considered and rejected by Jersey and Socony, and that negotiations were underway to solve the French problem by giving them flexibility to meet their requirements with an

---

[102] Memorandum, Eakens to Wilcox, February 14, 1947, file: "(Press) General," box 2, Records of the Petroleum Division, RG59.

[103] Memoranda, Nitze to Clayton, February 21, 1947, and Eakens to McGhee, March 4, 1947, *FR(1947)*, V, 646-51.

increased share of IPC oil.[104] In addition to all of this, Walter J. Levy, then an analyst in the Petroleum Division, undertook a thorough study of the history of IPC and the validity of the claims in the French note, and concluded that the United States would be on firm ground if it supported Jersey and Socony diplomatically.[105] On April 10, a formal reply was made to the French ambassador, expressing the hope that negotiations then in progress would result in C.F.P. being able to meet its needs from Iraq, declining any governmental obligation to support the 1928 agreements, expressing no opinion on the validity of the agreements under British law, and pointing out that restrictive clauses such as those contained in the agreements were incompatible with current American policy.[106] Stripped of formal phraseology, this was firm diplomatic support for the Aramco project.

Diplomatic support did not, however, solve the legal problem. Jersey and Socony had indeed considered resolving the matter by one buying out the other's interests in IPC and the other going into Aramco alone. Sheets, especially, favored the idea, but Jennings and Harden preferred to share the risk in both areas, and it proved impossible to determine a price for the IPC stock on which both companies could agree.[107] It was therefore decided to

[104] Memorandum of conversation, Harden, Jennings, Nitze, Loftus et al., undated, reporting a meeting on March 7, *FR(1947)*, V, 651-54; and memorandum on "Meeting . . . [at] . . . Department of State . . . March 7, 1947," U.S. Congress, *Multinational Corporations*, pt. 8, p. 160.

[105] Memorandum, Levy to Rayner, Loftus et al., March 6, 1947, unmarked red binder, box 6, Records of the Petroleum Division, RG59. Levy had begun his work on petroleum in the Office of Strategic Services during World War II, and subsequent to his service in the Department of State, he became a highly respected petroleum consultant.

[106] Letter, Acting Secretary of State to Bonnet, April 10, 1947, *FR(1947)*, 657-60.

[107] Letter, Johnson to Piesse, January 17, 1947, excerpt from minutes of Texas board meeting, January 31, 1947, letter, L. E. Hanson to Jennings et al., February 25, 1947, letter, Holton to Jennings, March 5, 1947, letter, Holton to Socony board, March 5, 1947, minutes of Socony board meeting of March 11, 1947, U.S. Congress, *Multinational Corporations*, pp. 130-34, 149-50, 156-59, and 163-66; memorandum of conversation, Harden, Jennings, Nitze, Loftus et al., undated, reporting a meeting on March 7, 1947, *FR(1948)*, V, 651-64; Case interview.

go ahead with the original plan. On March 12, 1947, a series of documents were signed consummating the sale, but contingent on satisfactory resolution of the litigation pending in the British courts. Under the agreements, Jersey and Socony together guaranteed a bank loan to Aramco for $102 million, which would be paid off as soon as Aramco could legally accept $76.5 million from Jersey and $25.5 from Socony in return for 30 and 10 percent interests in the business. This left Socal and Texas with 30 percent each. The money from the loan was used to repay Socal and Texas $79.8 million that had been advanced to Aramco, and to pay a dividend of $22.2 million to the original parents. In addition, Jersey and Socony became partners in Tapline and guaranteed their proportionate share of a $125 million loan for that project. A further agreement provided Socal and Texas with overriding payments out of Aramco's earnings over a number of years totaling $367.2 million, bringing the total amount received by Socal and Texas for a 40 percent interest in Aramco to approximately $469.2 million.[108] At least from the short-range financial point of view, Colliers and Rodgers had made a good deal—provided C.F.P. and Gulbenkian could be satisfied.

Negotiations with C.F.P. moved swiftly, and by May 1947, agreement had been reached that C.F.P. would withdraw its objections in return for the flexibility to increase its share of IPC production as required and a commitment by Jersey and Socony to support a sharply increased production program for IPC.[109] Final settlement was delayed, however, by difficulties with Gulbenkian, and to protect their interests Jersey and Socony filed a countersuit in the British courts in June asking that the 1928 agreements be declared "null and void and unenforceable in law."[110] Gulbenkian's problem stemmed from the fact that he had

[108] U.S. Congress, *International Petroleum Cartel*, pp. 119-128.

[109] Cable, Harden, and Sheets (London) to Jennings (New York), May 14, 1947, and press release, June 2, 1947, file 32929, and minutes of Jersey executive committee meetings of May 15 and 27, 1947, file 33369, FTC Records. The final settlement with C.F.P. and Gulbenkian is well described in U.S. Congress, *International Petroleum Cartel*, pp. 99-107.

[110] U.S. Congress, *International Petroleum Cartel*, p. 104; and minutes of Jersey executive committee meeting of October 8, 1947, file 33369, FTC Records.

no refining and marketing organization of his own, and his income was based on how his share of IPC production was handled. His leverage derived from the fact that under the 1928 agreements he could veto any changes, and he knew that the other companies were reluctant to have their affairs aired in court in order to have that agreement declared invalid. He was living in Lisbon at the time, and for over a year negotiators shuttled back and forth between New York, London, and Lisbon to try to reach an agreement. At one point, Gulbenkian told Case that he simply would not respect himself unless he "drove as good a bargain as possible."[111] He drove a good one. In return for dropping his objections to abrogation of the original agreement, he received rights to a sizable quantity of IPC production over and above his 5 percent share for the next fifteen years and complex guarantees that the oil would be marketed and that he would be paid in dollars.[112] The settlement was reached one day before arguments were due to begin in London on the suit and countersuit, and all parties barely missed having their affairs opened to the public in court. The final documents, which replaced the 1928 agreement as the operating bylaws of IPC, were signed November 3, 1948.[113] This cleared the way for the dropping of the C.F.P. suit and Jersey/Socony countersuit, and the consummation of the Aramco merger in December 1948.[114]

[111] Case interview.

[112] Memorandum of conversation, Harden, Nitze et al., December 22, 1948, FR(1948), V, 64-66.

[113] "1948 Documents," attached to letter, C. L. Harding to Nitze, December 16, 1948, 800.6363/12-1948, RG59.

[114] Minutes of Jersey executive committee meetings of November 18 and December 6, 1948, file 12750420-422 (pt. 2C), FTC Records.

VI

FORM A COALITION

WHILE corporate officers had been negotiating a merger in the Middle East, the world in which they lived had changed radically.[1] The total defeat of Germany and Japan, the devastation of France and Italy, and the exhaustion of Britain had left the United States and the Soviet Union as the surviving major world powers. The United States had emerged from the war economically powerful and in possession of the atomic bomb, but with its armed forces in the process of precipitous demobilization. Its ideological viewpoint was that of liberal democracy and modified laissez-faire economics, and its leadership believed that for the emerging world order to be a satisfactory one it would have to be hospitable to those ideas and their attendant institutions. The Soviet Union, on the other hand, emerged from the war economically debilitated, in possession of the largest land army in the Western world,[2] and

[1] Among the considerable body of writings on the origin of the Cold War, three are especially useful for the interpretations they offer: Daniel Yergin, *Shattered Peace: The Origins of the Cold War and the National Security State*; Louis J. Halle, *The Cold War as History*; and Bruce R. Kuniholm, *The Origins of the Cold War in the Near East: Great Power Conflict and Diplomacy in Iran, Turkey, and Greece*. For an interesting analysis of Stalin's role in this era, see Vjotech Mastny, *Russia's Road to the Cold War: Diplomacy, Warfare, and the Politics of Communism, 1941-1945*.

[2] The Soviet military establishment was a large one after World War II, but it was not as large as believed in most American government circles at the time. It was generally thought that the postwar Red Army would stabilize at 3.5 to 4.5 million men. In fact, it was reduced from 11.4 million in mid-1945 to a low point of 2.9 million in early 1948. This compared with a reduction in the total American military establishment from over 12 million in mid-1945 to 1.5 million in mid-1947; Yergin, p. 270; and John Lewis Gaddis, *The United States and the Origins of the Cold War, 1941-1947*, p. 261. The Chinese Nationalist Army in 1946 numbered approximately 3 million; *The China White Paper, August 1949*, vol. 1, p. 313.

in jealous occupation of Eastern Europe—the corridor through which it had been invaded in 1812, 1915, and 1941. Its political culture had been authoritarian through much of its history, and the latest incarnation was a ruthless, Marxist leadership that viewed the capitalist world with great distrust.[3] In retrospect, confrontation between nation-states in such different circumstances and with such different world views was almost inevitable, but as Daniel Yergin has pointed out, the precise form of the confrontation was by no means foreordained.[4] The form it took was the product of what Louis Halle has called a mutual self-fulfilling prophecy, with each side taking progressively stronger "defensive" measures that were viewed as "offensive" and responded to accordingly by the other,[5] until the pattern of reciprocal hostility was fully institutionalized by the Korean War in 1950. Within the United States this pattern translated into increasing anxiety over the security of the country and an increasingly militaristic policy of "containing" communism in general and the Soviet Union in particular.

It is important to keep this climate of opinion in mind because these events created an atmosphere that considerably magnified the strategic importance of the Middle East and Persian Gulf oil. Matters came to a head in February 1947, when Britain notified the United States that because of acute economic difficulties it could no longer provide aid to Greece and Turkey. On March 12 (by coincidence the same day that Jersey and Socony concluded the Aramco agreement) Truman went before a joint session of Congress to ask for $400 million in aid for those two countries. He did so in the context of a disturbing commentary on the world situation and a statement that the United States must "support free peoples who are resisting attempted subjugation by armed minorities or by outside pressures."[6] The Truman Doctrine, as this policy came to be called, was a major milestone in the postwar

[3] For an interesting essay on the Soviet view of this period, see Isaac Deutscher, *Stalin: A Political Biography*, pp. 571-604.

[4] Yergin, pp. 7-13.   [5] Halle, p. 148.

[6] Norman A. Graebner, *Ideas and Diplomacy: Readings in the Intellectual Tradition of American Foreign Policy*, p. 731.

replacement of Britain by the United States as the Western power most directly interested in strategic defense of the Middle East.[7] On June 5, worry over the economic and political viability of Europe led to Secretary of State Marshall's proposal of a comprehensive European recovery program aimed at "revival of a working economy . . . to permit the emergence of political and social conditions in which free institutions can exist."[8] Access to Persian Gulf oil for the European economy came to be viewed as critical to the success of the Marshall Plan.

As tension mounted, the American military became increasingly concerned that it lacked the capability to defend the far-flung perimeter defined as vital by the policy makers. The Berlin blockade and airlift of 1948 and 1949, the Soviet explosion of an atomic bomb in August 1949, and the fall of China to Mao Tse-tung in the autumn of 1949 did nothing to allay those fears, but the government had great difficulty in arriving at a coherent posture agreed to by all parties.[9] Under the National Security Act of 1948 (as amended in 1949), the armed forces were unified under a secretary of defense (initially Forrestal), the air force was made co-equal with the army and navy, military command was reorganized under the Joint Chiefs of Staff, and a National Security Council of key cabinet members was created to advise the president. However, each service continued to respond to the developing air of crisis in accordance with its own strategic doctrine. The army wanted universal military training and the capacity to mobilize four million men within one year; the navy wanted flush-

[7] Extensive intergovernmental discussions were held in Washington in October and November, 1947, to coordinate the British and American positions vis-à-vis the Middle East during what turned out to be the period of transition from British to American preeminence in the area. These are documented as "The Pentagon Talks of 1947" in *FR(1947)*, V, 488-626.

[8] Graebner, p. 733.

[9] This description of the military policy debate of the late 1940s is based on Samuel P. Huntington, *The Common Defense: Strategic Programs in National Politics*, pp. 25-64; and Warner R. Schilling, "The Politics of National Defense: Fiscal 1950," pp. 1-266, and Paul Y. Hammond, "NSC-68: Prologue to Rearmament," pp. 267-378, in Warner R. Schilling, Paul Y. Hammond, and Glenn H. Snyder, *Strategy, Politics, and Defense Budgets*.

deck supercarriers; and the air force wanted seventy combat-ready airgroups and the capability of delivering nuclear weapons to the Soviet Union itself.[10] Collectively these demands would have totaled $30 to $40 billion for the FY1950 budget that began to be considered in Congress in the fall of 1948, well in excess of the $14.4 billion ceiling that Truman considered politically feasible at the time.[11] The ensuing debate was intense, and was not resolved at a policy level until approval by Truman of NSC-68, a National Security Council policy paper prepared by an ad hoc State-Defense study group at the instigation of States' policy planning staff, in April 1950. NSC-68 defined national policy as rapid creation of adequate forces-in-being to deter the Soviet Union from aggression and "confront it with convicing evidence of the determination and ability of the free world to frustrate the Kremlin design of a world dominated by its will."[12] The doctorinal quarrel between the services was resolved with emphasis on the air force, since funds invested in its strategic nuclear delivery systems were deemed the most efficient method of achieving deterrent capability.[13] All that remained was the political will to fund a rearmament program, and that was generated by the North Korean invasion of South Korea in June 1950. Expenditures for national security rose to $22.3 billion in fiscal 1951, $44 billion in fiscal 1952, and $50.4 billion in fiscal 1953.[14] The significance of all of this for our purposes, however, was the fact that during the late 1940s concern over the security of the Persian Gulf ran well ahead of capacity to defend that area with forces-in-being. This discrepancy had a profound effect on the thinking of State and Defense Department officers concerned with Middle Eastern policy in the late 1940s.

[10] Huntington, p. 45.   [11] Ibid., pp. 43 and 50.

[12] NSC-68, "United States Objectives and Programs for National Security," April 14, 1950, copy held by the Modern Military Branch, Military Archives Division, National Archives, Washington, D.C., pp. 64-65. Truman's approval, in April, 1950, was by implication. He actually only referred the document to the National Security Council to spell out in more detail the specific programs and costs involved. But when the Korean War provided the incentive for rearmament, NSC-68 became de facto the basic policy guide.

[13] Schilling, p. 211.   [14] Huntington, p. 54.

The problem was nowhere more apparent than in the contingency planning conducted by the Joint Chiefs of Staff, beginning with studies under the code name "PINCHER" in December 1945.[15] By June 1946, a tentative strategic concept had been developed for a war with the Soviet Union, and it was strikingly similar to that employed against the Axis powers in World War II. The plan envisioned a "main offensive effort in Western Eurasia," and "active defensive [effort] in eastern Asia," holding key lines of communication (including the Mediterranean), and "maximum strategic air bombardment . . . against vital areas of the Western U.S.S.R."[16] As planning progressed, it became clear that the navy was the service most concerned with logistics for a long war, and therefore the one most concerned about defense of the Persian Gulf area.[17] But there was a limit to what even the navy could do. In early 1948 it was estimated that it would take at least *six divisions* to secure the area, but actual assignments of American forces then available to meet all of the planning requirements left only *one reinforced Marine battalion* to be detailed to Bahrain from Guam to assist in evacuation and denial opera-

[15] For an interesting discussion of the navy's postwar contingency planning for the Middle East, see David A. Rosenberg, "The U.S. Navy and the Problem of Oil in a Future War: The Outline of a Strategic Dilemma, 1945-1950," pp. 53-64. I am indebted to Professor Rosenberg for bringing to my attention the material on this subject in the records of the Office of the Chief of Naval Operations.

[16] Memorandum, "Résumé of PINCHER Planning," Cato D. Glover to Chief of Naval Operations, January 21, 1947, file: "A16-3(5) War Plans," OP30 Files, CNO Records.

[17] Memorandum, "Notes on the World Oil Situation," R. L. Dennison to Secretary of the Navy, October 17, 1946, file: "Politico-Military Policy Division, Subject Files (2), 1946-1947"; memorandum, ". . . Medium-Range Emergency Plan," Louis Denfeld to Joint Chiefs of Staff, April 5, 1948, file: "War Plans," OP30/A16-3(5) Files; memorandum, "The Navy Position Regarding the Unilateral Request of the Air Force for a 70 Group Program," C. W. Styer to Chief of Naval Operations, April 10, 1948, file: "Strategic Plans" OP30/A21 Files; memorandum, ". . . Study of the Nature of Warfare Within the Next Ten Years . . . ," W. F. Boone to Chairman of the General Board, May 17, 1948, file: "Warfare Reports," OP30/A16-3(R) Files; and memorandum, ". . . Alternative Short-Range Emergency War Plan," L. Denfeld to Joint Chiefs of Staff, September 22, 1948, file: "War Plans," OP30/A16-3(5) Files; CNO Records.

tions.[18] (Even this was eliminated in 1950 as a result of the budgetary emphasis on retaliatory air strikes.)[19] It was hoped that British Commonwealth forces would be available for the area, but this was by no means certain. So bleak was the picture in 1948 that the Joint Chiefs of Staff and the National Security Council began work on a detailed plan (NSC-26) for denial of the Saudi fields to the Soviets in the event of war. It called for advance preparations to destroy all above-ground facilities within a matter of hours, and to plug all wells with concrete blocks 1,000 feet down, which would take an estimated thirty days. The latter was designed to make the wells unusable without the waste of reserves caused by fire or dynamite.[20] There is no evidence that this plan was ever discussed with 'Abd al-'Aziz, or ever finally approved, but its consideration in 1948 was indicative of the limited forces then available for defense of the area.[21]

This capability was at total variance with the importance placed on Middle Eastern oil in concurrent planning for logistic support of a major war effort of extended duration. A full-scale Joint Logistics Committee study in early 1947 concluded that in "a

[18] Memorandum, ". . . Study of the Nature of Warfare Within the Next Ten Years . . . ," W. F. Boone to Chairman of the General Board, May 17, 1948, Enclosure H, p. 9, file: "Warfare (Reports)," OP30/A16-3(R) Files; and memorandum, "Tentative Assignment of Forces for Emergency Operations," C. W. Nimitz to Commander in Chief, U.S. Atlantic Fleet et al., July 12, 1948, file: "War Plans," OP30/A16-3(5) Files; CNO Records.

[19] Memorandum, "Joint Outline Emergency War Plan . . . ," S. H. Ingersoll to Commander in Chief, U.S. Naval Forces, Eastern Atlantic and Mediterranean, January 30, 1950, file: "Warfare Operations; War Games (1950)," OP30/A16-3 Files, CNO Records.

[20] J.C.S. 1833/1, "Preparations for Demolition of Oil Facilities in the Middle East," April 7, 1948, and SANACC 398/4 [same title], May 25, 1948, file: "CCS600.6 Middle East (1-26-46)," J.C.S. General Files, 1948, RG218; memorandum, George F. Kennan (director of the State Department's Policy Planning Staff) to [Undersecretary of State Robert A.] Lovett, August 23, 1948, memorandum, Sidney W. Souers (executive secretary, National Security Council), January 10, 1949, and memorandum, Kennan to Assistant Director, Office of Policy Coordination, CIA, May 11, 1949, file: "Strategic Materials-Oil," box 12, Records of the Policy Planning Staff, RG59.

[21] The final disposition of this plan remains a matter of conjecture; as late as 1979 the remainder of the files on this subject were still classified.

major future war of five years duration . . . the total United States
military and civilian consumption requirements . . . [could not]
. . . be met after M + 3 years by all of the then current production
in the United States and United States controlled foreign sources,
including that in the . . . Middle East, even with the . . . drilling,
new refinery, and synthetic plant building programs proposed.''
The study warned of growing dependence on Middle Eastern oil
as strategically risky, and it strongly recommended extensive preparation to produce ''synthetic petroleum from our vast domestic
reserves of natural gas, coal, and shales . . . as the safest and
most prolific means to . . . meet the demands of a future war.''
In the meantime, it called for conservation of ''domestic and
United States controlled foreign crude petroleum resources other
than in the . . . Middle East by maximum importation in peacetime
of crude petroleum from the . . . Middle East, consistent with the
maintenance of a healthy petroleum industry in the United States,
South America, and the East Indies.'' It pointed out that the
''Middle East could produce much more petroleum . . . with a
greater return for a given effort than . . . any other area,'' and
that ''Saudi Arabia . . . [was] . . . particularly valuable to the
United States'' because of the American concession there.[22] The
Department of State continued to share this basic viewpoint with
the military. An extended position paper in late 1948 pointed out
that it was ''essential that the development of . . . [Saudi oil]
. . . be allowed to continue and that the United States and other
friendly nations have access to it. For this to occur, it will be
necessary to keep the goodwill of the King and other important
Saudi Arabs and to prove to them that American business initiative
in developing the oil of Saudi Arabia in the best possible way for
the Government and people of the country.''[23] As Cold War concerns mounted, the position of the Departments of Defense and

[22] J.C.S. 1741, ''Problem of Procurement of Oil for a Major War,'' January
29, 1947, pp. 1-7, file: ''CCS463.7 (9-6-45) Sec. 6,'' J.C.S. Decimal File, 1946-
47, RG218.

[23] State Department position paper prepared in advance of the ''Pentagon talks''
with representatives of the British Government in October and November, 1947,
FR(1947), V, 553.

State vis-à-vis the Arab world became increasingly clear and consistent. Access to, and development of, Persian Gulf oil had become a vital national interest.

The major complicating factor, of course, was Palestine and the Zionist drive for a Jewish homeland. The roots of the issue ran deep. Palestine had been the principal home of the Jews from about 1,100 B.C. until their dispersal after a revolt against Roman rule in A.D. 132-35. For centuries the area was controlled by Byzantines, Persians, Arabs, Crusaders, Mongols, Egyptians, and, from 1517 until 1920, the Ottoman Turks. The population during all of this time was predominantly Palestinian Arab. By 1900 the Jewish population of Palestine was only about 50,000 out of a world Jewish population of 11 million,[24] but intense persecution—especially in Eastern Europe—had rekindled a desire to return to the Biblical homeland. This desire had both religious and secular dimensions and resulted in the founding of the World Zionist Organization in 1897 in Basle, Switzerland. By 1917 the movement had grown strong enough to elicit from British Foreign Secretary Lord Balfour a vague pledge to support the establishment of a Jewish home in Palestine, and when Britain acquired a mandate over Palestine from the League of Nations in 1920, the door was opened for Jewish immigration. At that point 90 percent of the total Palestinian population of 700,000 was Arab, and Jerusalem was the third most holy city in Islam, next to Mecca and Medina. By 1939 the influx of immigrants brought the Jewish population to about half a million (30 percent of the total) and provoked such strong Arab opposition that the British agreed to sharply limit further immigration. The death of 6 million Jews in the Holocaust, however, gave added force to the Zionist drive for a real and symbolic place of refuge, and the heavy influx of refugees from Nazi concentration camps as illegal immigrants into the British mandate created a highly volatile situation just after World War II. The Arabs, of course, were incensed at this invasion of the land of *their* ancestors, and protested vigorously and con-

[24] The statistics in this paragraph are from W[asif] F. Abboushi, *Political Systems of the Middle East in the 20th Century*, pp. 212-15.

sistently any British or American actions that appeared to yield to Zionist demands.

As already noted, the Departments of Defense and State in the late 1940s were basically in opposition to the movement for a Jewish state for reasons of national security, but overall American policy toward Zionism had been a vascillating one.[25] Both the Democratic and Republican platforms in 1944 had supported the idea of a Jewish homeland, but Roosevelt and the Department of State had repeatedly assured 'Abd al-'Aziz and other Arab leaders that no resolution of the Palestine question would be made without "full consultation with both Arabs and Jews."[26] However, just after the war, the United States government pressed the British to admit more Jewish refugees into Palestine, chiefly out of humanitarian concern for their immediate plight.[27] Then, in 1947, when Britain announced that it would relinguish its mandate and turn the matter over to the United Nations in 1948, the United States supported the UN decision to partition Palestine. The decision was vehemently opposed by the Arabs, and in March 1948, the United States withdrew its endorsement and proposed a temporary UN trusteeship instead. There was little enthusiasm for this idea in the UN, and it went nowhere. On May 14, 1948, the British withdrew, the Jews proclaimed a state of Israel, and full-scale war commenced with Arabs from Palestine and all of the

[25] For differing analyses of American policy during this period, see J[acob] C[oleman] Hurewitz, *The Struggle for Palestine*; Joseph B. Schechtman, *The United States and the Jewish State Movement: The Crucial Decade: 1939-1949*, which is based on an original study under the auspices of the Board for Research of the History of the Jewish People in Jerusalem; Herbert Feis, *The Birth of Israel: The Tousled Diplomatic Bed*; Richard P. Stevens, *American Zionism and U.S. Foreign Policy, 1942-1947*, which is highly critical of Zionist tactics; John Snetsinger, *Truman, the Jewish Vote, and the Creation of Israel*, which argues that the ultimate reason why Truman overrode State's objections and decided to recognize Israel in 1948 was consideration for the Jewish vote in an election year. Earlier brief, but useful, overviews include J. C. Hurewitz, *Middle East Dilemmas: The Background of United States Policy*, pp. 100-155; Nadav Safran, *The United States and Israel*, pp. 5-46; and William R. Polk, *The United States and the Arab World*, pp. 161-97.

[26] Cordell Hull, *The Memoirs of Cordell Hull*, 2, p. 1535.

[27] Harry S. Truman, *Memoirs*, 2, *Years of Trial and Hope*, pp. 132-42.

surrounding states. The United States reversed itself again and
formally recognized the state of Israel eleven minutes after it came
into being. The decision was made by Truman himself—two days
after he had received diametrically opposed advice from White
House political advisor Clark Clifford and Secretary of State
George C. Marshall and Undersecretary of State Robert A. Lovett.
With an eye to the New York vote in the upcoming presidential
election, Clifford argued for prompt recognition of what would
clearly be a fait accompli. Marshall and Lovett opposed the in-
trusion of domestic politics into foreign policy and argued against
such precipitous action on the grounds of national security.[28] The
decision elated the Zionists, dismayed the Arabs,[29] and provoked
impassioned debate for decades thereafter. For our purposes, however, the important point to note is that State remained opposed
to alienation of the Arab world down to the very end. American
recognition of Israel in 1948 ran counter to the well-articulated
views of State, Defense, Aramco, and 'Abd al-'Aziz.

One of the charges made at the time and subsequently was that
State's position was due to excessive corporate lobbying.[30] It is
true that, ever since 1937,[31] the companies had warned the department that American support for Zionism would alienate 'Abd
al-'Aziz and possibly result in "destruction of the U.S. political
and economic position in the Middle East . . . [and] . . . loss of
the oil concession in Saudi Arabia."[32] But by the late 1940s this

[28] Snetsinger, pp. 11-114.

[29] Letter, [Ambassador to Saudi Arabia] J. Rives Childs, June 24, 1948, file:
840.1, box 685, RG84. In reporting 'Abd al-'Aziz' dismay, Childs recommended
that if it were planned to continue American support for Israel, consideration
should be given to allowing Britain to assume responsibility for the Dhahran air
base, and to advising Aramco to shift its incorporation to Canada.

[30] American Zionist Emergency Council pamphlet, "Palestine Partition and
United States Security," February, 1948, 867N.01/2-1248, RG59. The inference
that State's position was the result of oil company lobbying is repeated in Schechtman, pp. 16-17.

[31] Memorandum of conversation, Moffett, Welles, Murray, and Feis, July 12,
1937, 890F.6363/Standard Oil Company/93, RG59.

[32] Memorandum of conversation, Duce, Henderson et al., November 4, 1946,
867N.01/11-446, RG59.

point was so abundantly clear that no special advocates were
required, and the companies adopted a relatively low profile on
the subject of Zionism itself.[33] By 1947 all of the key people in
State and Defense were aware of the strategic problem, the Soviet
threat, and the Arab position. Periodic Arab threats to cancel the
American concessions and impede the building of the pipeline
were more than adequately reported through diplomatic chan-
nels.[34] And there was little need for special lobbying when the
chief spokesmen within the government for nonalienation of the
Arabs were Secretary of Defense James V. Forrestal and Director
of the Office of Near Eastern and African Affairs Loy W. Henderson.

Forrestal's position dated back at least to 1943, when, as undersecretary of the navy, he began to stress the strategic impor-

[33] This low profile was a deliberate company policy, decided upon in part "for
fear of . . . recriminations which would have been leveled at oil interests." By
standing aloof from this volatile issue, the company also made it possible for the
Saudi government to differentiate to Arab critics between opposition to American
governmental actions vis-à-vis Palestine, and good business relations with the oil
companies; airgram, Tuck to Secretary of State, December 17, 1947, 890.6363/
12-1747, RG59. Active involvement in the Palestine debate would probably have
been counterproductive for Aramco both in the United States and in Saudi Arabia.
See also, memorandum of conversation, Duce, Henderson et al, May 28, 1948,
890F.6363/5-2848, RG59.

[34] Telegram, [Ambassador to Egypt] Tuck to Secretary of State, February 25,
1948, *FR(1948)*, V, 6-7; telegram, Tuck to Secretary of State, March 3, 1948,
890F.6363/3-348, and telegram, Childs to Secretary of State, June 10, 1948,
890F.6363/6-948, RG59. The Saudi Government itself was very clear about where
its interests lay and acted with great caution. In December, 1947, Prince Sa'ud
(son of 'Abd al-'Aziz) told Ambassador Childs that Iraq and Transjordan had
requested that Saudi Arabia break relations with the U.S. and cancel the oil
concession because of U.S. support for the Palestine partition plan. They had been
told that "Saudi Arabia was at one with other Arab states in opposition [to the]
establishment [of a] Jewish state but saw no reason [to] run counter to Saudi
Arabia's own interests by severing relations with [the] U.S."; telegram, Childs
to Secretary of State, December 15, 1947, *FR(1947)*, V, 1340-41. In May 1948,
however, Duce reported to Henderson that, for the first time, 'Abd al-'Aziz might
be forced by Arab public opinion to apply sanctions against the American concessions; memorandum, Henderson to Secretary of State, May 26, 1948, *FR(1948)*,
V, 15.

tance of American access to Saudi reserves. His position was completely consistent when in late 1947 and early 1948 he argued that Palestine policy should be de-politicized and decided on the basis of national security. This, of course, translated into opposition to the Zionist goals, and he was sharply criticized for this by those of opposite persuasion.[35] Even harsher criticism was meted out to Henderson, who was subsequently vilified for everything from anti-Semiticism to being a pawn for the oil companies and a spokesman for the British Foreign Office.[36] A carefully researched recent study, however, concludes that Henderson's views were those of a confirmed Cold Warrior, with a deep antipathy toward communism derived from eighteen years of service related to Moscow and Eastern Europe prior to being transferred to Iraq in 1943 and NEA in 1945.[37] He opposed anything that would alienate the Arab world and weaken support for America in an area vital to containment of the Soviet Union.[38] Such views were consistent with the interests of the oil companies and 'Abd al-'Aziz, but they derived from long-term strategic interests amplified by Cold War concerns, not from the work of corporate lobbyists.

Although corporate officers maintained a low profile on Zionism per se, they were not reticent about seeking governmental support for the Trans-Arabian Pipeline, which ran into difficulty over transit rights and steel allocations in exact parallel with the Palestine debate. The line was planned to run over 1,000 miles along the northern border of Saudi Arabia, across Transjordan, and then either through Palestine or through Syria and Lebanon to the

[35] Schechtman, pp. 415-22; and Walter Millis, ed., *The Forrestal Diaries*, pp. 188-89, 309-10, 323-24, 344-49, 356-65.

[36] Schechtman, pp. 409-13.

[37] Allen H. Podet, "Anti-Zionism in a Key United States Diplomat: Loy Henderson at the End of World War II," pp. 155-87. Daniel Yergin notes that Henderson was among the original foreign service officers stationed as observers at Riga, Latvia, in the 1920s who developed a strong distaste for Soviet behavior, and whose growing influence contributed to the hostility toward the Soviet Union that developed after World War II; Yergin, pp. 29, 35, 37-39.

[38] Interview with Ambassador Loy W. Henderson, in Washington, D.C., July 18, 1977.

Mediterranean. The company initially wanted to keep its options open on the site of the Mediterranean terminal, and it decided to negotiate transit agreements with all five countries.[39] Responsibility for these negotiations was assigned to Aramco attorney William J. Lenahan, and 'Abd al-'Aziz was kept fully informed about the entire project.[40] An agreement was signed with no difficulty with the British high commissioner for Palestine in Jerusalem on January 7, 1946.[41] The agreement did not call for payment of a transit tax, because IPC already had a pipeline terminating at Haifa with no transit tax, and under an Anglo-American agreement signed in 1924, American firms were entitled to at least equal treatment in the Palestine mandate.[42] In the case of Transjordan, the question of a transit tax proved to be a stumbling block. The British mandate was in the process of being dissolved, and negotiations were transferred in early 1946 from the British Colonial Office (which still had legal jurisdiction) to be government-to-be of King Abdullah, which wanted remuneration for use of its territory. The oil companies were amenable to some type of payment, but the Colonial Office registered strong objections on grounds of the precedent it would set in other areas. At this point, the Department of State entered the discussions and advised both the companies and the British government that it strongly favored inclusion of a transit tax in the agreement.[43] In

[39] Memorandum of conversation, Duce, Merriam et al., February 15, 1946, 867N.6363/2-1546, RG59.

[40] Letter, Duce to Richard Sanger (NEA), November 23, 1945, 867N.6363/11-2345, RG59.

[41] Letter, L[owell] C. Pinkerton (consul general in Jerusalem) to Secretary of State, January 11, 1946, 867N.6363/1-1146, RG59. A copy of the agreement, under date of January 17, 1946, is in file V52598, E1419/1419/31, FO/371.

[42] Convention between the United States and Great Britain relating to rights in Palestine, signed in London, December 3, 1924, FR(1924), II, 212; Shwadran, p. 333; Parliamentary Debates, 422, col. 1860, May 15, 1946.

[43] Memorandum, Loftus to Henderson, February 5, 1946, telegram, Byrnes to W. J. Gallman (chargé in London), March 16, 1946, telegrams, Gallman to Secretary of State, March 18 and 24, 1946, telegram, Acheson to Gallman, March 25, 1946, FR(1946), VII, 18-27; letter, Gallman to [Sir] R[obert] G. Howe (Foreign Office), April 3, 1946, V52598, E3069/1419/31, FO/371; and minutes of Foreign Office meeting of June 6, 1946, V52599, E5552/1419/31, FO/371.

a line of argument developed by John Loftus, now chief of the Petroleum Division, State took the position that ''the security of our oil investments in the Near East . . . protection against hostile internal and external forces and . . . goodwill . . . [toward] . . . American companies will be much enhanced if the various countries in that area participate directly in the economic benefits resulting from the development of local oil resources.''[44] The Colonial Office yielded; an agreement including an annual transit tax was signed in Amman on August 8, 1946;[45] and Loftus' line of reasoning became a cornerstone of American policy for years thereafter. The companies might quarrel over the *amount* of economic benefits to be shared, but in principle they were in complete accord.

An agreement with the newly independent Lebanese government was signed in Beirut on August 10, 1946, including a transit tax,[46] and an agreement with Saudi Arabia was signed in Jiddah on July 9, 1947, with no initial transit tax because of the benefits the government would derive from increased sales of Saudi oil.[47] The real obstacle then proved to be Syria. The Syrian government wanted guarantees that the pipeline would terminate at a Syrian port, and various factions within the country wanted a variety of additional economic benefits. After a year of frustrating negoti-

[44] Memorandum, Loftus to Henderson, February 5, 1946, *FR(1946)*, VII, 18-22.

[45] Letter, British Legation, Amman, to Ernest Bevin (British Foreign Secretary), August 12, 1946 (enclosing a copy of the agreement), V52599, E8186/1419/31, FO/371. The annual transit tax was 60,000 Palestinian pounds, equal to approximately $250,000 (Benjamin Shwadran, *The Middle East, Oil, and the Great Powers*, p. 333).

[46] Copy in file 890E.6363/8-2246, RG59. The amount was based on the quantity of oil crossing Lebanon, with a minimum annual payment of 20,000 pounds sterling (editorial note, *FR(1946)*, VII, 29-30); this was equal to about $180,000 (Shwadran, p. 333).

[47] Copy in file 890E.6363/8-1947, RG59. Aramco already had the right to construct a pipeline, but Tapline was a separate company and a new agreement was required. Although the agreement itself did not provide for a transit tax, Ambassador Childs reported that one could be assessed after fifteen years (telegram, Childs to Secretary of State, July 9, 1946, *FR(1946)*, VII, 662-63). Presumably this was an understanding reached separately.

ations, the Tapline board of directors in San Francisco directed
Lenahan on August 14, 1947 to deliver a letter to the Syrian
government stating that, if an agreement were not signed by August 30, negotiations would be terminated. By inference, this
meant that the pipeline would bypass Syria and go through Palestine. When word of this reached 'Abd al-'Aziz, he was incensed
both by the inference, and by the lack of company sensitivity in
delivering so crude an ultimatum to an Arab government that
would be under great internal stress not to yield. He therefore
applied overt pressure on the company to withdraw the letter, and
discrete pressure on the Syrian government to sign if the company
did so.[48] American Ambassador George Wadsworth also strongly
advised Syrian Prime Minister Jamil Mardam Bey to sign if Tapline would modify the letter.[49] As a result, the letter was withdrawn, and an agreement essentially identical to the one with
Lebanon was completed on September 1, 1947.[50] The same internal political pressures, however, delayed ratification by the
Syrian parliament, and when news of American support for the
UN Palestine partition plan broke in November, action on the
transit agreement bogged down completely. Tapline field parties
in Syria and Transjordan had to be evacuated because of anti-
American violence, and the companies began to fear that the
pipeline might not be constructed at all.[51]

Meanwhile, the project had run into difficulties in the United

[48] Telegram, [Minister to Syria and Lebanon George] Wadsworth to Secretary
of State, July 30, 1946, *FR(1946)*, VII, 29; letter, Moose (chargé in Damascus)
to Secretary of State, March 17, 1948, 890D.6363/3-1747, and letter, Pinkerton
(in Beirut) to Secretary of State, August 26, 1947, with enclosures, 890D.6363/
8-2647, RG59. The enclosures to the latter are copies of company letters and
cables documenting this entire episode. A copy of Lenahan's letter to the Syrian
government, dated August 21, 1947, is in file 890E.6363/8-2247, RG59.

[49] Letter, Robert B. Memminger (chargé at Damascus) to Secretary of State,
September 2, 1947, 890D.6363/9-247, RG59. Until January, 1947, Wadsworth
had been minister to Syria and Lebanon. He was vacationing in Lebanon when
he was called to Damascus by the prime minister for informal personal advice.

[50] Copy in file 890D.6363/9-247, RG59. See also telegram, September 2, 1947,
890D.6363/9-247, RG59.

[51] Memorandum of conversation, Sanger and Philip Kidd (Aramco), December
26, 1947, and n. 2, *FR(1947)*, V, 668.

States. The company had decided to commence work in Saudi Arabia even before the western terminus had been decided upon, a construction organization had begun to be assembled, and 335,000 tons of steel pipe and auxiliary equipment had been ordered for delivery at the rate of 45,000 tons a quarter beginning in November 1947. But there was a steel shortage in the United States, export controls had been extended longer than had been anticipated, and in July 1947, the Department of Commerce decided to hold up the necessary licenses while it studied the matter. Terry Duce immediately appealed to the Near East and Petroleum Divisions of the Department of State.[52] There was no question but that the pipeline would be more efficient than tankers to move oil to Europe, and there was no question but that increased oil to Europe would permit diversion of Venezuelan oil to the United States, which at that time was experiencing spot shortages.[53] The problem was competing domestic requirements, and the feeling on the part of the American independents that steel for Tapline was just another form of government assistance to their competitors.[54] State, however, took up the cause within the government, in part because, as Loy Henderson argued in a memorandum to Undersecretary Lovett, "the entire Arab world is exercised over the trend of American policy on the Palestine question. . . . In view of the publicity which the projected pipeline has received in the Near East, a decision having the effect of preventing or delaying its construction would be regarded by the Arab States as being linked with our 'unfriendly' Palestine policy. A beneficent economic policy in the Near East is one of the few means we have of offsetting . . . the effects of our Palestine policy."[55] State was backed initially by the navy on the familiar grounds that "an increase in production . . . [in the] . . . Middle East . . .

[52] Memorandum of conversation, Duce, Merriam (NEA), Hoffman (PED) et al., July 29, 1947, and memorandum, Hoffman to Eakens, September 3, 1947, 890F.6363/10-247, RG59.

[53] Memorandum, Henderson to Lovett, September 4, 1947, with enclosures, 890.6363/8-2847, RG59.

[54] *Oil and Gas Journal* 46 (October 4, 1947), p. 40.

[55] Memorandum, Henderson to Lovett, September 24, 1947, 690F.119/9-2447, RG59.

would relieve the drain on United States and Western Hemisphere reserves.''[56] After extensive interdepartmental reviews and final approval by the president's cabinet, the Commerce Department on September 26 issued the necessary fourth quarter licenses[57] and steel began to be shipped. The Independent Petroleum Association of America immediately charged that the steel would allow "thoughtless interests to further augment their strength in foreign oil fields,''[58] and Senator Kenneth Wherry's Small Business Committee on October 15 began a series of hearings on this and related subjects that lasted until July 10, 1948.[59] For the next year and a half, steel allocations for Tapline were under constant attack from critics in the United States.[60]

On June 18, 1948, at the peak of the first Arab-Israeli war, Senator Wherry succeeded in arranging an executive session with Secretary of State Marshall, Secretary of Defense Forrestal, Secretary of Commerce Charles W. Sawyer, and Assistant Secretary of the Navy Kenney to discuss the situation. In view of continued fighting in the area and the fact that Syria still had not granted transit rights, Wherry was able to obtain a commitment from all four that further export licenses for mainline pipe would be held in abeyance and no further licenses would be issued without prior notification of the Small Business Committee (until January 1, 1949, when the committee was due to expire).[61] Tapline continued a reduced level of construction in Saudi Arabia with steel already licensed, but when fighting subsided in Palestine, Duce came back to talk with Undersecretary of State Lovett about licenses for the

[56] Memorandum of conversation, David K. E. Bruce, undersecretary of commerce, W. John Kenney, assistant secretary of the navy, and thirteen others, September 22, 1947, 890F.6363/10-247, RG59.

[57] Department of Commerce announcement, September 26, 1947, 890F.6363/10-247, RG59.

[58] *Oil and Gas Journal* 46 (October 4, 1947), p. 40.

[59] U.S. Congress, Senate, *Special Committee to Study the Problems of Small Business, Hearings*, pts. 21-24, 25-28, 33-34, 36, 38, and 40-44.

[60] U.S. Congress, *Congressional Record, Senate*, March 11, 1949, vol. 95, pt. 2, pp. 2222-25, 81st Cong., 2d sess.

[61] Ibid., telegrams, Marshall to Legation to Saudi Arabia, June 19, 1948 (two), *FR(1948)*, V, 22-241.

remaining pipe. He emphasized the economic value of the project to Lebanon, Syria, Transjordan, and Saudi Arabia, and likened it to "a little 'Marshall Plan' for the Middle East, but . . . without cost to the American tax payer."[62]

Despite this obvious attempt to tailor arguments to the situation, NEA concluded that in "the recent setback suffered by all American interests in the Near East as a result of our stand on Palestine business firms have seemed to suffer less than either U.S. Government or American cultural interests. . . . It may well be . . . that the oil companies are in a position to recover lost ground . . . sooner than the U.S. Government."[63] By this time, the Department of Defense had concluded that the pipeline could not be supported as an immediate military necessity because of its previously noted indefensibility,[64] and State decided to emphasize economic and political factors. A letter went out over Marshall's signature to Secretary Sawyer supporting the license applications and arguing that the "oil of the Middle East . . . [was] . . . an important factor in the success of the European Recovery Program." In addition, completion of the project would "be of substantial benefit in relieving the demand on the oil reserves of the Western Hemisphere," and revenues from the pipeline would be a "stabilizing factor to the . . . economies of the countries in the area," helping to offset "disruptive tendencies conducive to the spread of communism."[65] Despite this support, further action on the licenses was deferred pending resolution of the problem with Syria. On February 24, 1949, when a transit agreement appeared imminent, the Department of Commerce resumed licensing.[66] In

[62] Letter, Duce to Lovett, August 25, 1948, attached to memorandum of conversation, Lovett, Duce et al., August 26, 1948, 890F.6363/8-2548, RG59.

[63] Memorandum, [Acting Director of Near Eastern and African Affairs Raymond A.] Hare to Lovett, August 25, 1948, *FR(1948)*, V, 39-40.

[64] National Security Resources Board memorandum of February 27, 1948, memorandum, Valentine B. Deale to Forrestal, March 19, 1948, and memorandum, Fleet Admiral William D. Leahy to Forrestal, March 19, 1948, file: "CCS678 (3/6/47) Sec. 1," J.C.S. General Files, 1946-47, RG218.

[65] Letter, Marshall to Sawyer, September 15, 1948, *FR(1948)*, V, 45-47.

[66] U.S. Congress, *Congressional Record, Senate*, March 11, 1949, vol. 95, pt. 2, p. 2223, 81st Cong., 2d sess.

March 1949, the Syrian government was overthrown in a coup
by Colonel Husni al-Za'im, who promptly signed an agreement
and saw to it that it was ratified along with his other actions by
a national referendum on June 26.[67] Work now commenced from
the western end, with the terminal located at Sidon, in Lebanon,
and transiting Syria and Transjordan to connect with the line in
Saudi Arabia. The final link was welded in September 1950, the
first oil reached Sidon in November, and the first tanker loaded
on December 2, 1950. The line was 1,068 miles long with an
initial capacity of 300,000 barrels per day, and the final cost was
$201.4 million, including pumping stations and the terminal at
Sidon.[68]

All of this requires little recapitulation. It was clear that with
the onset of the Cold War and the sale of a 40 percent interest
in Aramco to Jersey and Socony in 1948, a concert of interests
had coalesced around Saudi oil. The military wanted increased
production to conserve Western Hemisphere strategic reserves;
the Department of State wanted economic stability in the area to
guard against the spread of communism; Socal and Texas wanted
increased markets to secure the concession; Jersey and Socony
wanted additional sources of supply; and 'Abd al-'Aziz wanted
more revenue and modernization of his country. Truman's precipitous recognition of the state of Israel had the effect of increasing State's concern over relations with the Arab states, and
made State especially supportive of oil company projects such as
Tapline. The members of the coalition all saw Aramco as an
instrument for their own purposes, but it remained to be seen how
well the group would work when subjected to the internal stress
of differing opinions. Noticeably absent from the coalition were

[67] Memorandum of conversation, April 6, 1949, 890D.6363/4-649, telegram,
Keeley (Damascus) to Secretary of State, 890D.4-1449, and letter, Davies (Tapline) to Acheson, June 28, 1948, 890F.6363/6-2849, RG59. Colonel al-Za'im's
government lasted only five months (Abboushi, p. 307), but the succeeding military government honored the agreement. For a fuller discussion of the coup of
March, 1949, see Patrick Seale, *The Struggle for Syria: A Study of Post-war Arab
Politics, 1945-1958*, pp. 24-36.

[68] Henrietta M. Larson, Evelyn H. Knowlton, and Charles S. Popple, *New
Horizons 1927-1950*, p. 747.

the Federal Trade Commission, the Anti-Trust Division of the
Department of Justice, and nonparticipating American oil companies—especially the Texas independents. But their opposition
was muted in 1948.

*

THE FIRST real test of the coalition came in the fifty-fifty profit
sharing agreement of 1950, and the background to that decision
deserves careful examination. It was far more complex than conventional wisdom would have it and involved a corporate debate
on the price of Saudi oil, a British decision to exclude "dollar"
oil from sterling areas, a change in Venezuelan tax laws, the entry
of J. Paul Getty into Middle Eastern oil, and the reluctance of
'Abd al-'Aziz to adopt a budgetary system commensurate with
his new-found wealth.

First, there was the price of Saudi oil. It has already been noted
that the "price" charged by Aramco for crude and products transferred to the account of Caltex (the Socal and Texas marketing
subsidiary) was one arbitrarily set by the Aramco board.[69] In
March 1947, just before the merger, the transfer price for crude
stood at the relatively low figure of $1.02 per barrel f.o.b. Ras
Tanura, on the basis of cost (estimated at $.40 per barrel) plus
an amount designed to build up a cash and investment reserve
within Aramco.[70] This was essentially a bookkeeping matter of
little concern to anyone because profits made by Caltex or other
downstream subsidiaries on the sale of this oil ultimately wound
up on the books of the same parent corporations for distribution
in taxes and dividends. 'Abd al-'Aziz had no legal basis for com-
plaint, because his royalties were tied to production (at 4 shillings

[69] It was argued in the 1940s that Persian Gulf oil formed such a small part of
the world total that there was no adequate method of determining a separate market
value for it.

[70] Letter, C. E. Olmsted (Texas) to J. H. MacGaregill (Socal), March 22, 1947,
and letter, T. L. Lenzen (Socal) to MacGaregill, July 17, 1947, U.S. Congress,
Senate, Committee on Foreign Relations, Subcommittee on Multinational Corporations, *Multinational Corporations and United States Foreign Policy, Hearings*, pt. 8, pp. 176-78 and 194-95. Pt. 8 of these *Hearings* contains a large
number of corporate documents on this subject—obtained primarily from Socal.

a ton) rather than profits. It was clear to Socal and Texas, however, that an increase in the transfer price would increase the profit shown on Aramco's books and increase the risk that 'Abd al-'Aziz would demand a larger share in the yield from his own resources.[71] A 1947 Texas study further noted that over the period from 1947 to 1949 the $1.02 price would result in a yield of $44 million to the American government in income taxes from Aramco itself, compared with $65 million paid to the Saudi government in royalties.[72] A substantial increase in the transfer price would also change this ratio and create a situation that 'Abd al-'Aziz might consider grossly unfair.

Since this practice of establishing a low transfer price was essentially the same as that followed by Anglo-Iranian in Iran and IPC in Iraq,[73] Socal and Texas apparently assumed that Jersey and Socony would not question the practice in Saudi Arabia. Even if they did, Socal and Texas held a majority of the votes on the Aramco board of directors and could block any such action. They were due for a surprise. A quirk in the agreements made it considerably more profitable to Jersey and Socony to have the transfer price of crude raised to $1.48, a figure estimated to be still competitive with Venezuelan, West Texas and other Middle Eastern crudes when shipped to the Mediterranean.[74] The regular offtake agreement divided Aramco production on the same percentage basis as stock ownership (60 percent Caltex, 30 percent Jersey, and 10 percent Socony), but it would not become effective until Tapline was completed in 1950,[75] and an interim offtake agreement was signed to cover the years from 1947 to 1949. Since it would take some time to bring Aramco production up to the levels

[71] Memorandum, Lenzen to MacGaregill, August 2, 1947, U.S. Congress, *Multinational Corporations*, pt. 8, pp. 213-14.

[72] Letter, G. W. Orton to Rodgers et al., August 5, 1947, U.S. Congress, *Multinational Corporations*, pt. 8, pp. 215-16.

[73] Letter, Lenzen to MacGaregill, August 6, 1947, U.S. Congress, *Multinational Corporations*, pt. 8, pp. 219-21.

[74] Memorandum, J. A. Cogan (Jersey) to S. P. Coleman (Jersey), April 25, 1947, U.S. Congress, *Multinational Corporations*, pt. 8, pp. 180-81.

[75] Aramco Offtake Agreement-1947,'' U.S. Congress, *Multinational Corporations*, pt. 8, pp. 169-76.

now required, and since Socal and Texas did not want to limit
Caltex, the interim agreement set fixed quantities for 1947-1949
that worked out to 74 percent for Caltex and 26 percent for Jersey
and Socony combined.[76] On reflection, it is obvious that with
Aramco profits divided sixty-forty, Jersey and Socony would be
entitled to the Aramco profits on the extra 14 percent sold by
Caltex, and the higher the price, the larger the amount. At one
point, a Jersey study concluded that the difference between $1.02
and $1.48 converted into $14 million that would flow either to
Caltex or to Jersey and Socony, depending on how the issue was
settled.[77]

Socal and Texas understood this arithmetic, but they completely
overlooked an innocuous clause that Orville Harden had inserted
in the agreement.[78] The clause simply noted that Aramco ''should
be run for its own benefit as a separate entity,''[79] and after the
agreement was signed, Jersey obtained an outside legal opinion
that these words bound the Aramco board to make as much profit
as good business practice permitted, or be charged with dereliction
of duty.[80] Jersey and Socony therefore had both reason and leverage to press for a higher price.

The issue was joined at the very first meeting of the new Aramco
board, on March 12, 1947. The Jersey and Socony representatives
demanded a higher price, Socal and Texas refused to budge, and
the question was deferred for further study.[81] There then ensued
months of acrimony in correspondence, meetings, and threats of
legal action.[82] On November 18, Socal and Texas dug in their

[76] Letter, Olmsted to MacGaregill, March 22, 1947, U.S. Congress, *Multinational Corporations*, pt. 8, pp. 176-78.
[77] ''Exxon Paper-7/25/47,'' U.S. Congress, *Multinational Corporations*, pt. 8,
pp. 205-7.
[78] Follis interview.
[79] ''Aramco Offtake Agreement-1947,'' U.S. Congress, *Multinational Corporations*, pt. 8, pp. 169-76.
[80] Letter, Harden to Jennings, November 25, 1947, U.S. Congress, *Multinational Corporations*, pt. 8, pp. 237-38.
[81] Letter, S. P. Coleman to E. Holman and J. R. Suman (Jersey), March 13,
1947, U.S. Congress, *Multinational Corporations*, pt. 8, p. 168.
[82] Letter, Olmsted to MacGaregill, March 22, 1947; memorandum by G. V.

heels and Follis (in his capacity as chairman of the Aramco board) flatly informed Harden that the price would remain at $1.02 for the period March 12 through December 31, 1947.[83] That position held for the rest of 1947.

In early 1948, the situation began to change. In January, the Aramco board decided that it needed more funds for a five-year expansion program, and raised the transfer price to $1.30.[84] This was not enough to satisfy Jersey and Socony, and by coincidence or by plan they radically reduced their January liftings of Ras Tanura products, creating the threat of a refinery shutdown in late February. The stated reason was lack of tankers, but Socal noted that Jersey and Socony had not cut back production in Venezuela or the Netherlands Indies, and they saw this as a veiled threat.[85] The upshot was a decision to ignore the interim offtake agreement and let Jersey and Socony lift up to their pro rata 40 percent share of Aramco products and crude, beginning in January and February respectively.[86] This solved the original problem, but continued debate over exactly how the price should be determined finally led to a compromise in July 1948, setting a price that would compete with West Texas and Venezuela on the eastern seaboard of the United States. The figure decided upon for unstabilized crude at the July board meeting was $1.40 per barrel f.o.b. Ras

Holton on meeting of May 1, 1947; memorandum, W. L. Faust (Socony) to Sheets, July 31, 1947; memorandum, "Aramco Crude Price," September 10, 1947; letter, Follis to Harden, November 18, 1947; letter, Chester F. Smith (Jersey) to W. F. Moore (Aramco), December 4, 1947; letter, W. L. Faust (Socony) to Moore, December 4, 1947; letter, Harden to Collier, December 22, 1947; letter, Moore to Harden, January 27, 1948, U.S. Congress, *Multinational Corporations*, pt. 8, pp. 176-78, 182-83, 210-11, 223-27, 237-39, 241, and 246.

[83] Letter, Follis to Harden, November 18, 1948, U.S. Congress, *Multinational Corporations*, pt. 8, p. 237.

[84] Memorandum, Harden to G. Keegler (Jersey), January 26, 1948, U.S. Congress, *Multinational Corporations*, pt. 8, p. 246.

[85] Letter, Moore to Collier, February 7, 1948; and memorandum, Lenzen to MacGaregill, February 13, 1948; U.S. Congress, *Multinational Corporations*, pt. 8, pp. 247-48.

[86] Memorandum by A. C. Long (Texas), December 8, 1948, U.S. Congress, *Multinational Corporations*, pt. 8, p. 280.

Tanura.[87] For our purposes, however, the key point was the fact that this price resulted in a larger yield to the American government in taxes in 1949 than the yield to Saudi Arabia in royalties, and when the Saudi government learned of this, they used the differential as a major argument for a larger share in the profits from their own oil.[88]

As this point it should be noted that Senator Frank Church's Subcommittee on Multinational Corporations of the Senate Foreign Relations Committee reached quite a different conclusion on the episode just recounted. Its report, issued in 1975, argued that the reason Jersey and Socony fought for the higher price was to keep low-cost Saudi oil from undercutting them in the United States and Europe.[89] It is true, as the report points out, that Socony and Jersey appear to have discussed the possibility of doing this at least once in 1946, and it is true that avoidance of antitrust problems was one argument that Socal used in opposition to the price increase.[90] It is also true that a side effect of the price increase was a situation that made it unprofitable for any one of the four companies to radically cut its price for the American market and at the same time sell more than its ownership percentage of Aramco production. A court of law might find this in violation of antitrust law, but from the historian's viewpoint a careful reading of the Church Committee's own extensive collection of corporate documents on the subject makes it clear that the real issue under debate in 1947 was division of Aramco's profits and not curtailment of competition.

[87] Letter, Coleman to J. A. Cogan (Jersey), July 12, 1943; and memorandum, Coleman to J. W. Connally (Jersey), July 19, 1948, U.S. Congress, *Multinational Corporations*, pt. 8, p. 278.

[88] Taxes paid to the American government in 1949 were reported to have been about $40 million, compared with $39 million in royalties paid to the Saudi government; telegram, Childs to Secretary of State, June 13, 1950, *FR(1950)*, V, 52; and Table IV-4, above.

[89] U.S. Congress, *Multinational Oil Corporations and U.S. Foreign Policy*, pp. 77-81. Blair repeated this charge in his *The Control of Oil*, pp. 40-42.

[90] Letter, Jennings to Sheets, July 24, 1946; and memorandum, G. W. Orton (Texas) to Rodgers et al., August 5, 1947, U.S. Congress, *Multinational Corporations*, pt. 8, pp. 95-96 and 215-19.

Shortly after this price decision highlighted the inequities in
distribution of profits from the Aramco venture, another chain of
events raised the specter of slower growth in production. The
problem originated with the difficulties Britain experienced in
rebuilding it postwar economy, the resulting shortage of its dollar
earnings, the exchange restrictions it placed on purchases in dollars rather than sterling, and the effect this had on the ability of
Jersey and Socony to sell Saudi oil inside the sterling bloc. It will
be recalled that the possibility of such a development had been
alluded to by the British treasury and adamantly opposed by the
American delegation in the Anglo-American petroleum talks of
1944. But whether the Americans liked it or not, British economic
difficulties reached crisis proportions in the spring of 1949; the
pound was devalued to almost half of its former value; and severe
exchange restrictions were imposed. The immediate effect was
a drastic curtailment of the ability of Jersey and Socony to sell
Saudi oil (classified as "dollar" oil by the British) not only to
Britain, but to other countries in the sterling bloc—especially
Sweden, Denmark, Norway, and Finland. When the British also
resorted to barter arrangements to ease their exchange problems
and traded sterling oil for Argentine beef and various Egyptian
products, those two countries were added to the list. Because
postwar petroleum production also temporarily caught up with
postwar demand in 1949, the future for expanded production of
dollar oil as opposed to sterling oil in the Eastern Hemisphere
began to appear rather cloudy.[91]

The immediate impact in Saudi Arabia was a slowdown in
production below the figures originally targeted for 1949 and a
decision by the Aramco board to slow its expansion program from
seven drilling strings in the field in January to five in July 1949.[92]
It should be emphasized that production continued to climb from
the level of 1948, but not at the rate originally forecast. Since

[91] Larson, Knowlton, and Popple, pp. 701-6; Page interview

[92] "Working Paper, Near East Conference," December 20, 1949, file: "Near
East Oil," box 2, Records of the Petroleum Divison, RG59; and letter, McPherson
to Moore, June 22, 1949, U.S. Congress, *Multinational Corporations*, pt. 7, pp.
95-97.

'Abd al-'Aziz' royalties at the time were based on production, it was clear to all concerned that this slowdown would have an immediate impact on the rate of growth in the king's income. The slowdown also infringed on the fighting spirit of Aramco field personnel, who had not wanted to be tied to Jersey and Socony in the first place. Citing what he considered unagressive marketeering by Jersey and Socony and undue restrictions by the parents on field operations, Jim McPherson resigned as secretary-treasurer and senior resident officer and accepted a roughly comparable position with his old friend Ralph Davies' Aminol operation in its new concession in the Saudi-Kuwait Neutral Zone.[93]

Also profoundly disturbed by this turn of events was the Department of State, which looked upon the economic well-being of Saudi Arabia as an important element in its Cold War Middle Eastern policy. A departmental position paper in December 1949, pointed out that "the current dollar-sterling crisis in oil could reduce the Saudi Arabian annual income by 25 million dollars," or one-quarter of its anticipated level. It noted that "British controlled production in . . . Iraq, Iran, Kuwait, and Qatar" was increasing without restriction and warned that continuation of these trends "might stalemate Saudi . . . progress while neighboring states [especially Hashemite Iraq] advance, jeopardizing the unique cooperation and friendship . . . existing between . . . [the] U.S. and Saudi Arabia." It argued that the "Western orientation of Saudi Arabia, which counter[ed] Arab reaction to Western support of Israel, would suffer" and recalled that the "Saudi delegate was responsible for blocking Arab League sanctions against the Middle East oil industry during the Palestine War." In strategic terms, the paper pointed out that "Saudi Arabia has fewer Communists than any strategically located country in the world" and an economically healthy Aramco provided an "effective weapon against [the] advance of Communism by . . . [providing the] . . . local populations with a livelihood [and] programs for health, education [and] sanitation."[94] In short, the

[93] Letter, Parker T. Hart (Consul General at Dharhran) to Secretary of State, July 2, 1949, U.S. Congress, *Multinational Corporations*, pt. 7, pp. 85-89.

[94] "Working Paper, Near East Conference," December 20, 1949, file: "Near

dollar oil crisis awakened in the Department of State a renewed
interest in assuring 'Abd al-'Aziz a solid and growing income.

In December 1949, the British went one step further and established a requirement that sterling-area affiliates of American
companies had to purchase their oil from British and Dutch-owned
companies to the extent that those companies had surplus oil to
sell.[95] Admittedly, petroleum had become the largest single drain
on Britain's dollar reserves, but both Jersey and the Department
of State protested vehemently the arbitrary classification of all
Jersey oil as "dollar oil" and all Anglo-Iranian and Shell oil as
"sterling." In fact, Jersey argued, "the dollar and non-dollar
costs of petroleum operations of British and American companies
are, area by area, very similar."[96] In addition, the company protested that American aid funds were being used by British companies to build facilities that would help them displace American
companies in sterling areas.[97] This was too much. With full support from the Department of State, Economic Cooperation Administrator Paul G. Hoffman informed the British embassy in
Washington in early January 1950, that ECA funds would no
longer be made available for such a purpose, and if free dollars
were used in their place, ECA general funds would be reduced
in exact proportion.[98] With this strong backing from the American
government, Jersey's Howard Page finally negotiated a complicated but satisfactory settlement directly with the British treasury
in May 1950. The British agreed to end gasoline rationing, and
Jersey undertook to supply all of the additional gasoline required
by its British affiliates with payment in sterling. Instead of re-

East Oil," box 2, Records of the Petroleum Division, RG59. This was the basic
State Department position for the duration of the "dollar oil" crisis; memorandum,
Nitze to Secretary of State, April 27, 1949, 800.6363/4-2749, and memorandum,
Nitze to Secretary of State, May 12, 1949, 841.6363/5-1269, RG59; memorandum
of conversation, Childs with Truman, September 28, 1949, *FR(1949)*, VI, 1614-
15; and especially memorandum, Wilkins and Funkhouser (Office of African and
Near Eastern Affairs), March 15, 1950, *FR(1950)*, V, 34-35.

[95] Larson, Knowlton, and Popple, p. 706.
[96] Ibid., p. 708.   [97] Ibid., p. 709.
[98] Memorandum, Funkhouser to McGhee, January 9, 1950, *FR(1950)*, V, 9-
10.

mitting profits to the United States in dollars, Jersey would use the sterling proceeds to purchase needed goods and equipment manufactured in Britain. Along with a series of similar agreements worked out by Page in 1950 and early 1951, this arrangement essentially solved the dollar oil problem.[99] As noted above, however, the crisis had the effect of placing Saudi income high on the State Department's worry list all through the first half of 1950.

From the Saudi viewpoint it was becoming increasingly clear that income linked to production at a fixed (1933) royalty rate was considerably less profitable than would be income based on a division of profits. The idea on how to accomplish this was provided by Venezuela.[100] It will be recalled that in 1943 Venezuela had enacted a complicated set of tax laws that at the time were believed to have the effect of dividing profits equally with the oil companies. In practice, the laws did not produce the yield anticipated, and after the war, agitation against the oil companies built up steadily from the left wing in Venezuelan politics. Largely in response to this pressure, the ruling Acción Democrática party in November 1948 enacted an income-tax law that guaranteed a fifty-fifty division of profits. The law was a masterpiece of simplicity. All existing taxes and royalties were left intact, but it provided that if oil company income exceeded the aggregate of those taxes and royalties in a taxable year, the excess would be subject to an additional tax of 50 percent.[101] The change was less onerous to Jersey and its Creole subsidiary than might otherwise have been expected, because the tax could be taken as a credit against American income taxes, and money that would have been paid to the American government was simply paid to Venezuela instead. The new Venezuelan law was widely discussed in oil circles and fully known to the Saudi government.[102]

[99] Page interview; Larson, Knowlton, and Popple, p. 710-13; and telegram, Douglas (American ambassador in London) to Secretary of State, July 3, 1950, *FR(1950)*, V, 60-61.

[100] This discussion of the Venezuelan income-tax law of 1948 is based on Leonard M. Fanning, *Foreign Oil and the Free World*, pp. 71-110.

[101] Ibid., pp. 95-97.

[102] Memorandum, "Middle East Oil," by Richard Funkhouser, September,

A further inducement for change was provided by J. Paul Getty.[103] As already noted, Ralph Davies' American Independent Oil Company (Aminol) had acquired in June 1948 a concession from Sheikh Ahmad al-Subah of Kuwait for half-rights in the Kuwait-Saudi Neutral Zone. The price was high for 1948 (a royalty of $.33 a barrel compared with Aramco's $.22 per barrel), and instead of trying to match that on the Saudi side Aramco decided to relinquish its preferential rights there in return for Saudi recognition of a claim that its original concession included off-shore rights in the Saudi area of the Persian Gulf. As a result, the Saudi government was in a position to sell its half-rights in the Neutral Zone to the highest bidder. This turned out to be J. Paul Getty, whose Pacific Western Oil Company acquired the concession in January 1949 with an agreed royalty of $.55 per barrel, an initial payment of $9.5 million, an annual payment of $1 million, and Saudi rights to purchase shares in Pacific Western and to receive one-eighth of production profits and one-quarter of refinery profits. This was by far the most lucrative concession agreement in the Middle East at the time, and, if nothing else,

1950, *FR(1950)*, V, 76-96; and testimony of Ambassador George C. McGhee, U.S. Congress, *Multinational Corporations*, pt. 4, p. 85. The suggestion was made several years later in a Senate hearing that the Saudi government had been advised to enact the fifty-fifty tax law by a United States Treasury representative visiting Saudi Arabia in 1948. This was not true. The treasury representative in question was George Eddy, who was in Saudi Arabia along with Raymond Mikesell for six weeks to advise the Saudi government on monetary policy. Eddy was asked by a minor Saudi official to discuss "what kind of arrangements" there were "throughout the world by which the owners of oil-bearing lands participate in the income from the production of oil." Eddy cleared his response with the American ambassador and discussed the subject with the official once in very general terms. In the course of the discussion he did explain "the difference of the effect on the company between a royalty and an income tax," but he made no recommendations of any kind; U.S. Congress, Senate, Subcommittee of the Committee on the Judiciary and the Committee on Interior and Insular Affairs, *Emergency Oil Lift Program and Related Oil Problems, Joint Hearings*, pp. 1444-46.

[103] This discussion of Getty's entry into Middle Eastern oil is based on Longrigg, pp. 214-16. Aminol and Pacific Western drilled with no success through 1950, 1951, and 1952, and finally hit oil in significant quantities in 1953.

it convinced the Saudis that oil companies could afford to pay more than Aramco was paying under the terms of its 1933 concession. (To add insult to injury, it was an offer to head the joint field operation set up by Aminol and Pacific Western that lured Jim McPherson away from Aramco in the summer of 1949.)

All of these factors created an environment conducive to renegotiation of the concession agreement, but the actual impetus came from 'Abd al-'Aziz himself. He was a man of great wisdom and many virtues, but frugality and a concern for modern bookkeeping methods were not among them. His income had finally begun to rise, but it was continually exceeded by his generosity in tribal subsidies, modernization projects, and distributions to the royal family. As a result, his finance minister, 'abd Allah Al Sulaiman, was under continuous pressure to find the money to pay the bills. Aramco, in turn, was under continuous pressure from Al Sulaiman for loans, grants, and more generous interpretations of the existing agreements. The 1933 concession had provided for royalties at 4 shillings per ton, payable in British gold sovereigns, but it had not set a conversion rate for sovereigns into dollars. During and after the war Aramco began to pay partly in dollars, and from 1946 on 'abd Allah Al Sulaiman contended that the conversion rate should be that of the Jidda free market, which fluctuated between $16 and $20, rather than the official rate of $8.24. After long and arduous negotiation, Aramco and Al Sulaiman reached a compromise in 1948 and agreed to set the rate at $12 to the pound.[104] This had the practical effect of raising the royalty rate from $.22 to $.33 per barrel, and Aramco made a lump sum payment of $19.32 million to close out the controversy.[105] If the company thought that this would be the end of the matter, however, it was sadly mistaken.

Throughout 1949 and into 1950 Al Sulaiman continued to press Aramco for payment of the cost of harbor piers and railroads then under construction, contributions to a Saudi "welfare fund," royalties on oil used in refinery operations, deferred repayment of

[104] *FR(1950)*, V, 238, n. 2.
[105] Saudi Arabia, *Arbitration Between the Government of Saudi Arabia and Arabian American Oil Company*, I, 15, chart 8.

money already advanced, and—in general—a larger share in Aramco's profits.[106] As former Aramco General Counsel George W. Ray, Jr. recalled it later, "each time the company agreed to one thing, there was always just one more" to be discussed.[107] In July 1950, Moore as Aramco president, and Davies as executive vice-president went out to Saudi Arabia to discuss the situation with Al Sulaiman, and they took the position that if the company agreed to all of the demands, the resulting cost of Aramco oil would make it noncompetitive with others in the Middle East, to the disadvantage of Saudi Arabia. Al Sulaiman was not convinced, and he began to argue that the concession agreement itself should be renegotiated.[108] Under this kind of pressure, Aramco management began to cast around for some simple, straightforward formula that would put an end to the argument once and for all.[109]

In August 1950, the Aramco board finally authorized Moore to open negotiations with the Saudi government on possible revision of the 1933 concession agreement, with full knowledge that a change in the provision exempting Aramco from Saudi taxes would be the principal issue.[110] This was now late summer in

[106] Telegrams, Childs to Secretary of State, June 23 and July 25, 1950, *FR(1950)*, V, 58-60, 62-68. In addition to documents in the *Foreign Relations* series and the 1974 hearings on *Multinational Corporations*, this account of the negotiations that resulted in the fifty-fifty agreement is based on interviews with Robert I. Brougham, former financial vice-president of Aramco, by telephone from Cincinnati to La Jolla, California, December 2, 1977; George W. Ray, Jr., former general counsel of Aramco, by telephone form Cincinnati to East Thetford, Vermont, February 11, 1978; Douglas Erskine, former tax counsel for Aramco, by telephone from Cincinnati to Portola Valley, California, December 28, 1977; Ambassador George C. McGhee, former assistant secretary of state for Near Eastern, South Asian and African affairs in Washington, D.C., July 7, 1977; and George M. Bennsky, former assistant treasury representative in the Middle East, by telephone from Cincinnati to Washington, D.C., January 24, 1978; and letter to the author from Ambassador Richard Funkhouser, former petroleum advisor to McGhee, November 7, 1977.

[107] Ray interview.
[108] Telegram, Childs to Secretary of State, July 25, 1950, *FR(1950)*, V, 62-68.
[109] Ray interview.
[110] Telegram, Secretary of State to embassy in London, August 31, 1950, *FR(1950)*, V, 75-76.

1950. The Korean War had broken out, and NSC-68 had become the de facto policy guide within the Departments of Defense and State in matters relating to the Cold War. At the same time Aramco was preparing to renegotiate its concession agreement, Assistant Secretary of State for Near Eastern, South Asian, and African Affairs George C. McGhee was taking steps to shore up the American position in his area of responsibility.

A position paper prepared by Richard Funkhouser for McGhee in early September clearly delineated State's position. It noted that with "the threat of Communist aggression increasing throughout the world, the Middle East . . . [was] . . . highly attractive to the USSR because of oil, its strategic location . . . [and] . . . its vulnerability to attack from without and within." Economic progress, political stability, and Western orientation within the area were all critical to the containment of Communism, and the oil companies were well positioned to contribute to this objective. The most critical current issues, Funkhouser noted, were the financial terms of the concession agreements, and a number of Middle Eastern states were now pressing for agreements similar to "the Venezuelan sharing of profits arrangement." Failure to yield to such pressures might result in "mistakes like the Mexican expropriation," but in view of the shift in tax revenue away from the U.S. treasury that would result, State would be ill advised to actively press for such agreements. On the other hand, since "company retreat . . . [was] . . . inevitable, it would seem useful to make the retreat as beneficial and orderly as possible to all concerned." Whatever terms were finally agreed upon to increase the share of Middle Eastern states in the revenue from their own resources, the broad objective continued to be "the progressive development of Middle Eastern resources in order to preserve Western Hemisphere reserves . . . [and] . . . maintain political stability and economic progress in the Middle East."[111]

[111] Memorandum, "Middle East Oil," September, 1950, *FR(1950)*, V, 76-96. The position paper was prepared as background for a confidential meeting on September 11, 1950, between McGhee, Funkhouser, Colonel Drake, Terry Duce, E. L. De Golyer, Charles Harding of IPC and a number of others, in preparation for a trip by McGhee to London to discuss Middle Eastern policy; letter, McGhee

On November 6 an Aramco delegation called at the Department of State to confer on a response to the Saudi demands. The delegation included Fred Davies, Terry Duce, and Colonel William Eddy (now an Aramco advisor). They met with McGhee, Funkhouser, Loftus, Ambassador Childs, and several others. McGhee took the position that the department itself had no effective reason or means to oppose Saudi demands for an increased share in Aramco's profits and agreed that some change in the concession agreement was appropriate. The Aramco representatives were by now thoroughly convinced that their best avenue of retreat was along the lines of the Venezuelan fifty-fifty arrangement, but on the question of Aramco's eligibility for a foreign tax credit in this case, McGhee would take no position. That, he said, was a legal question "which could only be handled by the Treasury." Duce pointed out that—although he felt confident the tax credit would be granted—"the Bureau of Internal Revenue would not take a position on a theoretical case." The meeting ended with tacit agreement that the best course of action would be retreat along the lines of the Venezuelan agreement, even though there was no way to assure a tax credit in advance.[112] The next step was to convince the parent companies.

On November 13, Davies, Duce, and Moore arranged for a meeting between McGhee and his State Department team and senior officers of the parent companies—including Follis, Harden, and Jennings—to discuss the issue. The real purpose of the meeting, according to McGhee's memory years later, was to convince the parents to go along with what Aramco and the Department

to Drake, August 31, 1950, U.S. Congress, *Multinational Corporations*, pt. 8, p. 341. An account of the meeting itself is given in memorandum, Funkhouser to McGhee, September 18, 1950, U.S. Congress, *Multinational Corporations*, pt. 8, pp. 341-45. Funkhouser himself wanted increased competition by other American companies in Middle Eastern oil, but the corporate representatives present at the September 11 meeting were understandably unenthusiastic about the idea.

[112] Memorandum of conversation, McGhee, Davies et al., November (6?), 1950, *FR(1950)*, V, 106-9; Brougham interview; McGhee testimony, U.S. Congress, *Multinational Corporations*, pt. 4, pp. 85-92.

of State had already decided was the best course of action.[113] By this time, the decision-making process had been overtaken by events. Saudi Arabia had exercised its sovereign rights and decreed an initial 20 percent income tax on all individuals and companies within its jurisdiction (excepting those in the royal family, armed forces, religious posts, with annual incomes less than 20,000 riyals, or who paid the *zakat* religious tax—exceptions that had the practical effect of exempting all Arabs).[114] The companies had not yet decided how to respond to this move. Follis especially was concerned "that such arbitrary unilateral action would create a precedent which struck at the heart of all company contracts throughout the world," but Jennings noted that there was a body of legal opinion that "the terms of a concession which granted exemption from Saudi Arabian taxes . . . [were] . . . invalid on the basis that a sovereign cannot sign away his sovereign rights." McGhee repeated much of what he had covered at the previous meeting and said that the "Department agreed with the Aramco evaluation of the necessity of negotiating changes in their contract in the present circumstances." The rest of the meeting was a discussion of the various settlements that might be made. Davies stated that "from a psychological point of view . . . [the Venezuelan] . . . formula sounded fair and would be considered fair in Saudi Arabia." No formal decision was reached, but the meeting broke up with the understanding that this was the course that would be taken—still with no advance assurances on the tax credit issue.[115] In effect, Aramco had used State's Cold War concerns to convince the parents to yield to 'Abd al-'Aziz' pressure for additional income, and State had used the situation to accomplish one of its policy objectives in the Middle East.

Davies, Aramco General Counsel George W. Ray, Jr., and Financial Vice-President Robert I. Brougham flew out to Jiddah

[113] McGhee interview; and McGhee testimony, U.S. Congress, *Multinational Corporations*, pt. 4, p. 91.

[114] Saudi Arabian Royal Decree of November 4, 1950, U.S. Congress, *Multinational Corporations*, pt. 8, pp. 374-77; Brougham interview.

[115] Memorandum of conversation, McGhee, Harden et al., November 13, 1950, U.S. Congress, *Multinational Corporations*, pt. 8, pp. 345-48.

and opened negotiations with Al Sulaiman and Prince Feisal on November 28.[116] Al Sulaiman had a long list of additional demands that he was reluctant to give up, but Davies took the position that the company would agree to fifty-fifty profit sharing only if the Saudis would drop their other demands.[117] After a month of negotiations and consultations with the king, an agreement was finally signed on December 30, referencing a royal tax decree issued three days earlier.[118] The decree followed the Venezuelan formula, providing that all "companies engaged in the production of petroleum" would be required to pay "an income tax of fifty per cent (50%) of the net operating income." All existing royalties and taxes were left in place, but counted as credits against the Saudi income tax. Since the issue of an American tax credit had not yet been resolved, Aramco prevailed upon the Saudi government to include a provision permitting foreign (i.e., American) taxes to be deducted from operating income before the 50 percent profit sharing provision was invoked.[119] The Aramco-Saudi agreement of December 30 cleared up a number of outstanding items, but its key provision was one under which Aramco submitted to the income tax, anything in its concession agreement notwithstanding, "it being understood" that in no case would the taxes exceed fifty percent of Aramco's net income after deduction of operating expenses and foreign taxes.[120] Taken together, these two documents constituted the "fifty-fifty agreement of 1950."

At this point, it is necessary to clear away a further misconception about the agreement. The 1975 report of the Church Committee, and John Blair's *Control of Oil* both infer that prior to the agreement being consummated, the Treasury Department, on the advice of the National Security Council, made a special decision

---

[116] *FR(1950)*, V, 118, n. 1.

[117] Telegram, Ambassador to Saudi Arabia [Raymond A.] Hare to Secretary of State, December 12, 1950, *FR(1950)*, V, 119-20.

[118] Telegram, Hare to Secretary of State, December 31, 1950, *FR(1950)*, V, 121.

[119] Saudi Arabian Royal Decree of December 27, 1950, U.S. Congress, *Multinational Corporations*, pt. 8, pp. 377-78.

[120] "1950 Agreement," U.S. Congress, *Multinational Corporations*, pt. 8, pp. 372-74.

to let Aramco deduct its Saudi tax from its American income tax, as a device to subsidize the Saudi government at the expense of the American taxpayer without submitting the question to Congress.[121] This is simply not true. The decision to grant the tax credit was not actually made by the Internal Revenue Service until 1955, when Aramco's tax return for 1950 was finally audited.[122]

[121] U.S. Congress, *Report on Multinational Oil Corporations*, p. 85; and Blair, pp. 196-99. In both cases the inference is drawn from an exchange that took place between Senator Church and Ambassador McGhee during the course of the hearings (U.S. Congress, *Multinational Corporations*, pt. 4, pp. 84-95). The Senator charged, without benefit of documentation, that "upon the recommendation of the National Security Council, the Treasury made the decision to permit Aramco to treat royalties paid to Saudi Arabia as though they were taxes." Ambassador McGhee denied any first-hand knowledge of how the treasury came to make its decision, but went on to defend it as perfectly justified on grounds of national security. Neither party apparently realized at the time that the tax decision was made as a routine interpretation of the law five years after the fifty-fifty agreement. The Church Committee's own documents (U.S. Congress, *Multinational Corporations*, pt. 8, pp. 350-78) show that the Senator's allegations were incorrect, but both the authors of the report and Blair chose to use the exchange to infer National Security Council involvement and special treatment for Aramco. This incorrect inference in the committee *Report* is unfortunate, because it has led others to treat National Security Council involvement as fact; see, for example, Krasner, pp. 205-13, and Turner, pp. 47-48.

[122] Bureau of Internal Revenue Ruling 55-296, May 16, 1955, no. 20, U.S. Congress, *Multinational Corporations*, pt. 8, p. 358. As already noted, the Bureau of Internal Revenue would not make advance rulings on hypothetical cases. IRS practice at the time was to accumulate several years of corporate tax returns and audit them all at one time. If the local (in this case, New York) auditor had any questions on interpretation of the law, he would refer them to Washington for a ruling. In this case, the only question was whether Aramco's acquiescence in negating the provision of its 1933 agreement exempting it from Saudi taxes made the money an "agreed payment" equivalent to a royalty rather than a tax. The counter argument was that a sovereign power could not contract away its taxing power, and the tax exemption in the 1933 agreement had been invalid in the first place. When the company's 1950 tax return came up for audit in 1955, Aramco tax counsel Douglas Erskine gave the New York auditor copies of the relevant documents and heard nothing further from IRS until a favorable ruling was issued from Washington in May, 1955. In the interim (1953), the Saudis renegotiated the 1950 agreement to delete foreign (American) taxes from the deductions to be made from gross income in calculating profits. Had the 1955 IRS ruling gone the other way, Saudi revenue would therefore not have been affected. The only

As attested to by a 1957 report by the staff of the Joint Congressional Committee on Internal Revenue, the decision was a perfectly correct interpretation of a law that had been in effect since 1918.[123] Furthermore, two searches of National Security Council records by its Freedom of Information Office failed to produce any evidence that the tax question was ever considered by that body.[124] It is true that the Department of State solidly endorsed the fifty-fifty profit sharing decision itself for strategic reasons, and it is true that the workings of the long-standing law on foreign tax credits had the effect of transferring revenue from the American to the Saudi government, but the implication that the Internal Revenue Code was bent out of shape by the National Security Council to accomplish that objective is simply not correct.

To sum up, the sequence of events that led to the fifty-fifty decision provided an excellent example of the working of the coalition that had formed around Saudi oil. All parties wanted an increase in Saudi production and a continuation of the link with the United States, but within that framework, each party had its own particular ax to grind. The quarrel between Jersey and Socony on the one hand and Socal and Texas on the other over the transfer price for Saudi oil helped to highlight the inequities in the division of profits. When British exchange-restrictions threatened to slow Saudi production, Jersey and the Department of State reacted forcefully, and the debate had the effect of increasing State's Cold War concern over the adequacy of 'Abd al-'Aziz' income. Concurrent actions taken by Venezuela and J. Paul Getty convinced the Saudis that more income was possible from the Aramco concession, and the king's continuing generosity placed consid-

remaining question in 1955 was how much Aramco would pay in American taxes, and this issue was decided strictly on the basis of existing tax law; Erskine and Brougham interviews.

[123] Report by the staff of the Joint Committee on Internal Revenue Taxation (undated), U.S. Congress, *Multinational Corporations*, pt. 8, pp. 350-78. The date of this report is given as 1957 in U.S. Congress, Senate, Committee on Foreign Relations, Subcommittee on Multinational Corporations, *Multinational Oil Corporations and U.S. Foreign Policy, Report*, p. 91.

[124] Letter to the author from Beverly Zweiben, director, Freedom of Information, National Security Council, May 18, 1979.

erable pressure on his finance minister to obtain it. Aramco's own management served as the conduit for this pressure into the inner circles of the Department of State and the parent companies. State, in turn, used the situation to accomplish one of its own objectives and to strengthen the American position in a strategically critical area. The interactions apparent in this episode are a microcosm of American actions vis-à-vis Saudi Arabian and Middle Eastern oil for several decades thereafter. The point here is that this pattern had become firmly established by 1950 and is nowhere more evident than in the events leading to the fifty-fifty agreement of that year.

CONCLUSIONS

BEFORE any broad conclusions are drawn from the story just re-counted, comment is in order on several lesser points. The first concerns the original Socal concession in Saudi Arabia and is an excellent example of Anderson's law of perversity: That if things can turn out exactly opposite of what was originally intended, they probably will. The case in point, of course, is the Red Line Agreement of 1928. The clause in the agreement prohibiting the partners from operating independently within the confines of the old Ottoman Empire was intended by Gulbenkian and the French to coopt the Americans and avoid unwelcome competition in Middle Eastern oil. It was the price Jersey and Socony had to pay to gain access to Iraqi oil. But as matters turned out, the most significant consequence of the Red Line Agreement was a chain of events that brought two *more* American companies—Socal and Texas—into the Middle East and diverted Gulf into the far more prolific field in Kuwait. The agreement actually contributed to an increase rather than a decrease in the competition from American companies. It did provide Gulbenkian and Compagnie Française des Pétroles leverage for a better settlement with Jersey and Socony when those companies decided to get out in 1947, but that was about all. Ironically, the amount of rhetoric that has been generated by the Red Line Agreement has been totally out of proportion to the amount of oil actually produced by the Iraq Petroleum Company. As can be seen from Table I-2, Iraq accounted for only 1.9 percent of non-Communist world production in 1938 and 1.4 percent in 1950.

Aramco itself was an interesting and unique enterprise. The company evolved out of Socal's original field team, and it became an effective vehicle for corporate profit, the modernization of Saudi Arabia, and defense of American strategic interests in the

Middle East. Much of the credit for Aramco's success goes to the field personnel who set it on its original course, but they could do this only because 'Abd al-'Aziz had absolute control of his country, a determination to modernize it, and great skill in encouraging technical innovation within the framework of extreme Islamic conservatism. Cultural accommodation was facilitated by the historical accident of a long hiatus during World War II, which gave the "Hundred Men" an excellent opportunity to learn Arab ways. This learning time enhanced their ability to keep friction at a minimum when the influx of construction workers began in late 1944. But there was one more factor. There was something in Bedouin culture that made Saudi Arabs and American oil men surprisingly compatible on a personal basis. As one close observer put it, "Saudis and Americans laugh at exactly the same place in jokes."[1] All of these factors taken together contributed to a close and effective relationship between Aramco and the Saudi government, and the company became the centerpiece for a coalition of interests that coalesced around Saudi oil.

The Aramco venture became a matter of importance to the American government in 1943, when the Petroleum Administration for War, the navy, and the Department of State realized that the United States would soon shift from net exporter to net importer, and that an incredibly valuable source of supply was under concession to an American company in Saudi Arabia. By late 1943 State and Navy had concluded that it was in the highest national interest that America retain access to those resources, and that they be developed as rapidly as possible in order to conserve Western Hemisphere strategic reserves. Although the companies provided the government with data on the subject, the government's basic posture toward Saudi oil derived from the emerging pattern of world resources and not from corporate lobbying. At issue within the government was the means by which those objectives could best be achieved, and not the objectives themselves.

[1] Wanda M. Jablonski, editor of the *Petroleum Intelligence Weekly*, in a discussion of this subject with the author in New York on May 3, 1977.

One approach was to establish within the government the Petroleum Reserves Corporation, and through it to buy a controlling
government interest in Aramco itself and build a government-
owned pipeline across the Arabian Peninsula. The driving force
behind this approach was Harold Ickes, with his desire to dominate
American oil policy, but it was a poorly conceived scheme from
the beginning. In theory the PRC was modeled after British participation in Anglo-Iranian, and Ickes assumed that the basic objective could best be achieved by protecting the Aramco concession from British encroachment. But he misread the situation. In
the midst of World War II there was little real possibility that the
British would or could usurp the concession, and it is not clear
what the American government could have done to protect the
concession as a part-owner that it could not have done anyway.
The real way to protect the concession was to get Saudi oil onto
the world market, and the PRC's only contribution to that objective would have been helping Socal and Texas finance a refinery
and pipeline in Saudi Arabia. Socal and Texas were quite willing
to receive such aid, but the nonparticipating American companies
rose up in arms. As we have seen, Jersey and Socony blocked
the stock-purchase plan, and the Texas independents led the atttack
that effectively killed the pipeline idea. The real legacy of the
PRC was intense opposition to government-in-business from
the industry, Congress, and the state of Texas, which contributed
heavily to the demise of a far better idea—the Anglo-American
Petroleum Agreement.

In retrospect, the Anglo-American Agreement was intended as
a government cartel—exactly as charged by its opponents. Advocates of rational international planning in the development of
world resources could argue with some justification that it might
have been an experiment worth trying. If the International Petroleum Commission had developed into some type of deliberative
body representing government, oil companies, consumer interests,
and producers (as originally contemplated by the Department of
State) it is conceivable that the agonies of the last three decades
might have been prevented—or at least minimized. But the idea
ran completely counter to the American laissez-faire tradition, and

too many domestic interests saw themselves as adversely affected. In the midst of the firestorm created by the PRC in late 1944, the agreement foundered on opposition from the Texas independents. When the majors realized that they could not get the antitrust protection they wanted, they lost interest in the idea. And when State concluded that an agreement strong enough to accomplish its objectives was no longer possible, it, too, lost interest. The agreement was allowed to linger on to 1947 and to die a quiet death.

The effective demise of the agreement in early 1945 and the arrival on the scene of Will Clayton as a guiding force in economic policy resulted in a significant change in State's approach to the problem of Saudi oil. Along with the navy and the Joint Chiefs of Staff, State's objectives remained ensuring American access to, and development of, Saudi reserves in order to slow the drain on Western Hemisphere strategic resources. But it now turned to reliance on private enterprise to achieve these objectives. This approach dovetailed nicely with the interests of the oil companies and 'Abd al-'Aziz. Socal and Texas continued to worry about adequate markets for Saudi oil and money to build the Trans-Arabian Pipeline. Jersey and Socony were still supply-short and looking for ways to increase their participation in Middle Eastern reserves. And 'Abd al-'Aziz continued to press for an income commensurate with his concept of royal responsibilities. With full endorsement of State and Navy, Socal and Texas therefore proposed a merger of interests to Jersey and Socony, and their offer was promptly accepted. It took two years before the Red Line Agreement could be ended, but when it was in 1948 Aramco became a jointly owned subsidiary of all four companies. The de facto coalition now included the Department of State (backed by the navy and the Joint Chiefs), Socal, Texas, Jersey, Socony, Aramco itself, and 'Abd al-'Aziz. Each for its own reasons wanted increased Saudi production and retention of the concession in American hands. Absent from the coalition were the Anti-Trust Division of the Department of Justice and the other American oil companies—notably the Texas independents, but as already noted, their voices were muted in 1948. Jersey and Socony had been coopted into the coalition and State was now back on safe grounds by not supporting one domestic interest over another.

With the advent of the Cold War, the interest of the American government in Saudi Arabia deepened and took on a slightly different cast. 'Abd al-'Aziz was now seen as a key link in the defensive arc being built around the Soviet Union, and his economic well-being and good will became a matter of vital national interest to the Department of State, the Department of Defense, and the National Security Council. It was this viewpoint that created State's opposition to precipitous recognition of the State of Israel, support of steel allocations for Tapline, and encouragement of the fifty-fifty profit sharing agreement of 1950. All of these drew sharp criticism from opposing interests, but State's posture was at least consistent with its long-term view that Saudi oil ought to be developed to shore up the Middle East against Communism, and to slow the drain on Western Hemisphere reserves.

Although the low cost of Middle Eastern oil and other commercial factors were undoubtedly more influential than government policy, the statistics in Table I-2 suggest that events did move in the direction State intended. In 1938 the Eastern Hemisphere accounted for 11.5 percent of noncommunist world crude oil production, compared with 88.5 percent for the Western Hemisphere. By 1965 this ratio had shifted to 46.4 percent for the Eastern Hemisphere, compared with 53.6 percent for the Western Hemisphere. The only way this ratio could have been significantly improved would have been to curtail domestic American production and encourage the importation of then cheap Middle Eastern oil. But such action was far beyond the power of State (or any other government agency) in the 1950s, and the domestic opposition it would have aroused would have been formidable.

This study has not focused on the antitrust issues that have dominated so much of the writing on international oil, but no summation would be complete without a comment on that subject. As already noted, the Webb-Pomerene Act of 1918 permitted combinations overseas so long as they did not work to restrain trade within the United States. The intent was to make American companies at least as competitive as those from other countries, which did not share the American belief in the efficacy of antitrust. The problem, of course, was determining whether combinations overseas actually worked to restrain trade within the United States. This study suggests that the Aramco merger in 1948 did not in-and-of-itself create a violation of American antitrust law. It created a framework for possible collusion, but in-and-of-itself it was consistent with the intent of the Webb-Pomerene Act. The price debate between Socal/Texas and Jersey/Socony in 1948 was another matter. Although from the historical viewpoint the issue was division of profits rather than restraint of competition, the outcome was a situation that did discourage any of the partners from dumping low-cost Saudi oil in the United States in competition with the others. Whether a court would find this in violation of antitrust law is a matter for speculation, but from the author's point of view it was certainly on the fringe. Comment on other linkages and episodes involving the Seven Sisters would appear to be outside the scope of this study and best dealt with elsewhere.

*

IN RETROSPECT, Aramco was the centerpiece of an episode that suggests a great deal about the manner in which America's foreign oil policy has been formulated and executed. The fact that the United States had a position in one of the most prolific of the world's oil fields was a result of general American commercial expansion, the post World War I scare over domestic reserves that left Standard of California still prospecting in the Middle East in 1933, and a certain amount of pure historical accident. But when first established, this position was of real interest only to a handful of Socal geologists who were convinced that oil was there. In 1943, when the nation awoke to the fact that it would soon cease to be self-sufficient in oil and that an American company held a major concession in the Middle East, Saudi oil became a matter of vital national interest. The Saudis had no objection, because oil revenue, American technical know-how, and the protection of a major power in the postwar world were all highly to be desired.

The de facto American national policy toward Saudi oil that

evolved over the next seven years was a quite rational one, considering the conceptualization of the problem that was widely held in the government and the corporate community at the time. It was perceived as clearly in the national economic and strategic self-interest—as well as consistent with American laissez-faire ideology—to have Saudi oil developed as rapidly as possible by American private commercial interests, and this was the course taken. But this policy was not one arrived at by strong executive leadership (as the Constitution suggests that it should be), or by political debate within a duly elected legislative body. Instead, it was the product of competing interest groups in and out of government finally reaching a compromise and resolving themselves into a coalition that appeared to serve each of their special interests. As should be abundantly clear from the events of late 1944 and early 1945, the government agencies involved were the weaker partners in the coalition,[2] at least when it came to deciding on questions of method.

A similar pattern emerged in the Iranian crisis that followed close on the heels of the events recounted here. The crisis began when the fiery Iranian prime minister, Muhammed Mossedegh

[2] This observation is consistent with those of Peter J. Katzenstein, in a recent introduction to a set of comparative studies on the foreign economic policies of six modern industrial nations. He argues that "In their political strategies and domestic structures, the United States, Britain, West Germany, Italy, France, and Japan fall into three distinct groups. . . . [T]he Anglo-Saxon states rely, by and large, on a limited number of policy instruments which affect the entire economy rather than particular sectors or firms Policy makers in Japan, on the other hand, can pursue their objective of economic growth with a formidable set of policy instruments which infringe on particular sectors of the economy and individual firms. . . . In the two Anglo-Saxon countries the coalition between business and the state is relatively unfavorable to state officials and the policy network linking the public with the private sector is relatively fragmented. In Japan, on the other hand, state officials held a very prominent position in their relations with the business community and the policy network is tightly integrated." Germany, Italy, and France fall in the middle, with Germany and Italy closer to the United States and Britain, and France closer to Japan These differences in domestic structure have tended to be translated into foreign economic policies that were basically liberal in the case of the Anglo-Saxon countries, and neo-mercantilist in the case of Japan; Peter J. Katzenstein, ed., *Between Power and Plenty: Foreign Economic Policies of Advanced Industrial States*, pp. 20-21.

seized effective control of the government and nationalized the holdings of Anglo-Iranian in that country.[3] All of the other major oil companies invoked an embargo on Iranian oil until the issue was settled, and the resulting economic chaos in Iran seriously threatened the stability of a key link in Cold War defenses around the Soviet Union. State and Defense called upon the American majors to collaborate in handling the disrupted supply patterns to Western Europe and in evolving a solution that would put Iranian oil back on stream. But the companies demurred on the grounds that such collaboration would further jeopardize their position in criminal antitrust proceedings then being prepared by the Anti-Trust Division of the Department of Justice because of the network of cross linkages derived from the Iraq Petroleum Company, Aramco, Caltex, the Standard Vacuum Oil Company (a joint subsidiary of Jersey and Socony) and numerous long-term supply contracts with each other. On the recommendation of State, Defense, and the National Security Council, President Truman and then President Dwight D. Eisenhower intervened with the Department of Justice. The cases were reduced to civil proceedings and ultimately settled out of court, and the majors joined in an international consortium to market Iranian oil. The details of this complex sequence of events are far beyond the scope of this study, but the pattern of interacting interest groups was remarkably similar to that surrounding Saudi oil in the 1940s.

The same pattern was evident when import quotas were imposed by the United States on foreign oil in the late 1950s, although the effect of that decision was the exact opposite of the de facto policy on Saudi oil followed from 1943 onward. By the late 1950s low-cost Middle Eastern and other oil had begun to threaten the domestic industry (especially the Texas independents) and pressure was brought to bear on the executive branch to limit imports. The stated reason was again "national security," but this time the argument was to ensure survival of a healthy domestic industry. For all practical purposes, the objective of conserving Western

[3] For a thoroughly researched and carefully written account of American response to events in Iran in the early 1950s, see Burton I. Kaufman, *The Oil Cartel Case: A Documentary Study of Antitrust Activity in the Cold War Era.*

Hemisphere strategic reserves was abandoned when mandatory import quotas were imposed in 1959.[4] Again, the de facto national policy was the product of competing domestic interest groups rather than executive (or any other) leadership.

Much has occurred since the events recounted here. The foreign oil business was so lucrative in the 1950s that an increasing number of "independents" took up concessions in other areas; world productive capacity began to exceed world demand; prices declined; and in 1960 a number of nations joined together in the Organization of Petroleum Producing Countries to see what could be done to collectively ensure a better return on their resources.[5] The power to do this began to emerge in 1970, when American productive capacity peaked, dependence on OPEC oil increased significantly, and the consuming nations lost the capacity to simply embargo oil from a recalcitrant producer, as they had in the case of Iran in the early 1950s. By 1973, it was clear that effective control over supply and prices had shifted to OPEC, and the sequence of events that followed is well known. The United States plunged into a great national debate over how to respond to this new situation, and the end is not yet in sight. What is clear, however, is that the process is one of competing interest groups attempting to protect their own positions as a national concensus is hammered out. The debate surrounding Aramco and Saudi oil in the 1940s was an early microcosm of this same process.

One can only speculate on the outcome if an international petroleum commission had developed along the lines envisioned by the Petroleum Division of the Department of State in 1943. In theory it would ultimately have been a deliberative body with representation from the governments of both consuming and producing countries, with the power to make semi-binding recommendations on the production and distribution of much of the world's petroleum. If such a commission had proven itself capable

[4] Kaufman, pp. 72-75.

[5] For a thorough analysis of the events leading up to the Arab embargo on American oil in 1973, see Robert B. Krueger, *The United States and International Oil: A Report for the Federal Energy Administration on U.S. Firms and Government Policy* and Raymond Vernon, ed., *The Oil Crisis*.

of reconciling all of the conflicting interests that would have undoubtedly been brought to bear, it might have kept prices high
enough to avoid overdependence on petroleum, assured a substantial income to the producing countries, and kept the oil companies' return on investment at what would have been perceived
generally to be an equitable level, thus avoiding much of the
turmoil of the 1970s. But for such a commission to even have
had a chance of success, the United States would have had to
cooperate fully, and given the realities of the American political
and economic system described in this study, such cooperation
would have been very difficult to achieve. Unfortunately, the
commission never even had a chance to be tried in 1944 because
of the firestorm of opposition to government-in-business created
by Ickes and the Petroleum Reserves Corporation. If such an
international instrumentality is attempted in the future, one hopes
its architects will be able to take note of and avoid the shoals on
which the Anglo-American Agreement ran aground.

APPENDIX A

METHODOLOGICAL NOTE

OVER THE YEARS I have become convinced that differences in historical interpretation derive largely from differences in the conceptualizations of human society and the historical process that historians bring to their work. As Karl Mannheim pointed out long ago, different world-views focus attention on different aspects of complex events and may therefore lead to radically different conclusions.[1] Believing this to be true, I have developed a strong preference for explicit statements of underlying assumptions. Such statements place a work in intellectual context, clarify the perceptual screen through which the data has passed, and permit a critique of assumptions and method as well as conclusions. What follows, therefore, is a brief description of the theoretical framework and analytical method on which this study is based.[2]

*

TO BEGIN WITH, international relations have been conceptualized as the product of a single, complex, interlocked social system rather than of autonomous nation-states competing in a sea of anarchy.[3] Individual

[1] Karl Mannheim, *Ideology and Utopia· An Introduction to the Sociology of Knowledge*, p. xxi The citation is from the preface by Louis Wirth.

[2] I am indebted to Merton S. Krause, Eileen Bagus, James A. Stever, and Alfred Kuhn for having read and commented on an earlier draft of this Appendix. They do not agree with many of my views, but their criticisms have been most helpful in tightening the line of argument presented here. This conceptual model is far less rigorous than the one developed by Kuhn himself, but I found his *The Logic of Social Systems· A Unified, Deductive, System-Based Approach to Social Science* to be a *tour de force* in deductive reasoning.

[3] Arend Lijphart has argued that this world-system viewpoint constitutes a new paradigm that has emerged in thinking about international relations in the past few decades—replacing the older view of sovereign nations competing in anarchy See his "The Structure of the Theoretical Revolution in International Relations," *International Relations Quarterly* 18, no. 1 (March, 1974), pp 41-74, and "In-

nation-states are subsystems of the larger social system, and they in turn are made up of their own subsystems (or competing interest groups). National governments are subsystems of nation-states, not synonomous terms, and a clear distinction should be made between the policy of the "national government" and the policy of the "nation-state" (which is the de facto cumulative thrust of all competing interest groups within a nation-state extended outside of the national boundary). Furthermore, many subsystems cross national boundaries and operate in response to interests that may or may not coincide with those of a particular nation-state (multinational corporations, the European Common Market, and the Roman Catholic Church, to cite only a few). This conceptualization creates a rather complex model with which to work, but it is one that so far is undoubtedly shared by many contemporary historians.

The next step is one on which there may be less agreement.[4] For analytical purposes, I have designated each of the units in this vast system of multileveled, interlocked subsystems as "entities" that are either institutionalized groups, or (at the lowest level) individual persons. Entities are defined as either individuals, or groups with a perception by its members that they share in the destiny of that group and a formal or informal organization to make that destiny a satisfactory one (Standard Oil of California, the United States Navy, and the state of Texas, for example). Obviously, some groups will be more cohesive than others, there will be considerable overlap, and one person may belong to many groups. But the model assumes that each of these entities (individual and group) have exactly analogous characteristics.

The core characteristic is a drive for *self-actualization*, or a drive to

ternational Relations Theory: Great Debates and Lesser Debates," *International Social Science Journal* 26, no. 1 (April, 1974), pp. 11-21. Lijphart's concept of a "paradigm shift" is drawn from Thomas S. Kuhn, *The Structure of Scientific Revolutions*.

[4] Readers concerned with such matters may be interested in knowing that the metaphysical viewpoint that underlies this conceptual model is closely akin to the thinking of Pierre Teilhard de Chardin (*The Phenomenon of Man*), Baruch Spinoza (*The Ethics of Spinoza*, edited by Dagobert D. Runes, p. 27), and Georg Wilhelm Friedrich Hegel (*The Philosophy of History*). It includes the view that matter and spirit are simply different manifestations of the same phenomenon, that everything is connected with everything else, and that human history is the product of evolutionary directional change through a dialectical process, rather than recurring cycles or meaningless chaos. The conceptual model, however, should be seen as standing alone. It could be arrived at by a variety of other lines of reasoning.

become fully what it is conceived of as possible to become.[5] Actual motivation becomes rather complex, but in simple terms this means that navies want to be in a position to win wars at sea, and politicians want to win elections. These drives obviously conflict, but the interaction is a dialectical one, and larger cooperative groupings develop as perceptions of shared self-interest emerge. The classic example has been the rise of nationalism, which has bound men together in common cause without obliterating competing subgroups. The exact analogy may be applied to corporate mergers.

In addition to a drive toward self-actualization, entities have *belief systems* that are common to at least the dominant members of the group. The belief system provides a cognitive structure for organizing information, selecting a course of action, and deriving a sense of identity. Thus, the raison d'être of a business corporation may be long-term financial profitability, and this becomes at least one criterion for reaction to events in the outside world. And a corporate officer may see himself in part as a person who gains material and *psychic* benefit from taking action that brings profit to the corporation. A challenge to a corporation thus may become a challenge to the sense of identity of corporate officers who have both a material and a psychic stake in the survival of the enterprise. Exactly analogous functions derive from the dominant belief systems of the United States Air Force, the Communist Party, or the royal family of Saudi Arabia. Marxists argue that such belief systems emerge solely in support of what is perceived as necessary for the material well-being of a particular class, but that view is too narrow. I would argue that belief systems are coextensive with the perceived material *and psychic* well-being of any group, of which social class is only one. The belief system coexists with the group and is simply its ideated dimension. What complicates matters is the fact that a given individual belongs to many groups, and it is sometimes difficult to determine which is most powerful in a given situation. Despite this complexity, it is important to identify the dominant set of beliefs in each group under study, because the belief system will directly affect the course of action taken.

A third characteristic of entities (both individuals and groups) is power, or, more precisely, *relative power*. Power is defined here as the ability of A to cause B to do something B would not otherwise have done, or conversely, the inability of B to do something A does not wish B to do. There are obviously many interlocked dimensions of power, but four

[5] For individuals, this drive is well described in Abraham Maslow's "hierarchy of needs" in his *Motivation and Personality*, pp. 80-106.

stand out. The first is coercive power (military power externally and police power domestically)—the ability of A to bring effective force to bear on B. This of course varies not only with arms, training, motivation, and leadership, but also with the credibility of the threat to use force in a given situation. Another is economic power, defined here not only as financial power, but also as control over resources and technology that may be brought to bear by A to influence B. A third is political power, or the degree to which A can persuade others to support his cause in opposition to B. This support may result from psychic as well as material benefits that adherents see themselves deriving from A's cause. And the fourth is legal power, or the leverage A may derive over B as the result of the sanction of rules made by an institution to which both belong. This includes the leverage within bureaucratic organizations derived from rules that assign a particular responsibility to A rather than to B. Possession of coercive and economic power is relatively clear-out, and the sources of legal power are fairly easy to trace. It is political power, as defined here, that is most difficult to weigh.

The relative power that one entity has over another at each decision point should be carefully studied, because this determines which will prevail in a given situation. Before one recoils at the Nietzschean[6] tone of this statement, it should be noted that, under the definition of political power used here, an appeal to reason or to moral law may gain staunch supporters from those who derive psychic benefit from their identity as rationalists or as moral men. Thus, behavior that may be seen as altruistic in a material sense, may be in pure self-interest in the psychic sense. This is the source of much of the power in missionary and reform movements and accounts for the strong idealist streak in American foreign policy. Political power is most effective, of course, when supporters derive *both* psychic *and* material benefit from the cause.

A fourth characteristic of entities is that they exist in a *spatial context*. This is not an argument for geographic determinism, but it is obvious that geographic factors have played a major role in history. Consider the fact that in the mid-nineteenth century the projected canal across the Isthmus of Suez would considerably reduce the sailing time between Britain and her colonies in the East, or the fact that oil-poor Japan saw adequate reserves in the Dutch East Indies when cut off from American supplies in the summer of 1941, and saw the American fleet at Pearl Harbor as the major threat to a supply line from Sumatra to Honshu if

---

[6] Friedrich Nietzsche, *The Will to Power*.

those reserves were seized. In both cases, an examination of the spatial relationships involved explains a great deal about each nation's course of action.

A fifth characteristic of each entity is its own *internal dynamics*, which means that the interactions of *its* subsystems can be also analyzed in terms of *their* belief systems, relative power and spatial context. At some point, and by some reasonable criterion, the historian must decide how far down into the organization the analysis is to be carried.

Finally, it should be noted that entities usually have *specific objectives* in any given situation. Theoretically, if belief system, relative power, spatial context, and internal dynamics were fully understood, the objectives would be easily deductible. But since full understanding has thus far eluded the minds of men, this is not the case. A frequent problem arises because of the difference between what the participants in a given episode actually sought, and what we, from a different perspective in time, space, and culture, think they logically would have sought. A further complication derives from assuming that if a particular result followed from an episode, its attainment was the objective sought; this is frequently *not* the case. For these reasons, it is essential to search the record for the actual objectives of key participants.

*

BEFORE MOVING ON to the analytical method derived from this conceptual model, it is important to inject an additional comment. If this model is a reasonably accurate way of looking at human society, it is obvious that the causal chains, taken as a whole, are far too vast to be encompassed by a single mind, much less a single book. It is therefore necessary to enter the system at a carefully selected point, and to attempt a description of a particular event, or events, in terms of those entities most directly involved. All historians do this when they choose, for example, to write the biography of a confederate general, an account of a Civil War battle, or a social history of the South. The point here is that the choice of that portion of whole system on which to concentrate is critical and may significantly influence the conclusions reached. The story of the three blind men who felt the trunk, leg, and tail of an elephant and reached different conclusions about the nature of elephants is too well known to need amplification.

In terms of diplomatic history, this point has been well made by Graham T. Allison, who analyzes the Cuban missile crisis at two separate levels—the government as a national unity (his Model I), and the government as a conglomerate of competing bureaucratic groups (his Model III).[7] He demonstrates that alternative analytic approaches produce quite different conclusions but argues that both have value in attempting to explain the behavior of nation-states. The particular facts that one notices and comments upon as important may be quite different depending upon whether one thinks of the nation-state primarily as a single entity or as a collection of competing subsystems. In the case of this study, I have chosen to conduct the analysis on the basis of competing subsystems one or more levels below the nation-as-a-whole, but I have attempted conclusions at both that level and at the level of the nation-as-a-whole.

*

THE FOREGOING OBSERVATIONS lead directly to the analytical approach that was actually used. Its key elements are not startlingly different from those used by most historians, but the intent here is to make the process explicit.

The first step was phrasing of a basic research question, in order to focus the collection of data in relevant directions. In this case, the original question was: "What was the role of Aramco in the formulation and execution of American Middle Eastern policy during and immediately after World War II?" As research progressed, the question was reformulated to: "What was the role of each major participant in the process by which the United States focused its attention on Saudi oil and acquired a special relationship with Saudi Arabia in the 1930s and 1940s?" This focused attention at least one level below the nation-state as a single entity, and concentrated on the interaction of the subentities most deeply involved.

[7] Graham T. Allison, *Essence of Decision: Explaining the Cuban Missile Crisis*. An earlier, and quite interesting, discussion of different conceptualizations of the nature of international politics from the perspective of political theory is Kenneth N. Walz, "Man, The State, and the State System in Theories of the Causes of War." Walz examines theories of political behavior that focus on human nature (Spinoza), on the internal structure of nation-states (Kant), or on the anarchical nation-state system (Rousseau), and points out that none by itself offers a completely satisfactory explanation of war. Like Allison, he concludes that insights may be gained by examining international behavior from more than one perspective. For another example of this phenomenon, see my "Lend Lease for Saudi Arabia: A Comment on Alternative Conceptualizations," *Diplomatic History* 3, no. 4 (Fall 1979), pp. 412-23. In addition to Allison, another excellent study of the bureaucratic process in foreign policy is Morton H. Halperin, *Bureaucratic Politics and Foreign Policy*.

The reformulation led directly to subquestions on the role of the major participants in each of the following episodes:

1. The original acquisition of a concession in Saudi Arabia by an American company. (Chapter I)
2. The abortive attempt by the American government during World War II to buy an interest in the company holding that concession. (Chapter II)
3. The abortive attempt by the government to negotiate a petroleum agreement with the British during World War II. (Chapter III)
4. The interaction between Aramco and the Saudi government in Saudi Arabia. (Chapter IV)
5. The postwar linkage of Jersey and Socony with Socal and Texaco in the ownership of Aramco. (Chapter V)
6. The role of the government in supporting the Tapline project; the role of Aramco and its parent companies in the formulation of a policy toward Palestine; and role of all participants in the fifty-fifty profit sharing agreement of 1950. (Chapter VI)

The next step was to identify the "operative entities" that appeared to have interacted to produce the outcome in each episode. This, of course, depended heavily on subjective judgment. The approach taken was to identify the key decision points in each of the above episodes and to determine what entities (individuals or groups) most directly influenced those decisions. The list, arranged in outline form as subentities, is as follows:

1. The United States (an emerging power)
    a. The United States government
       (1) The executive branch
           (a) The president (Franklin Roosevelt; then Harry Truman)
           (b) The defense establishment (the Army-Navy Petroleum Board; the navy; the joint chiefs of staff; and James Forrestal)
           (c) The Department of State (Petroleum Division; Office of Near Eastern and African Affairs, and Will Clayton)
           (d) Harold Ickes (who dominated the Department of Interior, the Petroleum Administration for War, and the Petroleum Reserves Corporation)
           (e) The Department of Justice (Anti-Trust Division)
           (f) The National Security Council (in the late 1940s)
       (2) The legislative branch
           (a) Senate (Foreign Relations Committee)
    b. Oil Companies

    (1) Texas-based independents
    (2) Nonparticipating majors (including Jersey and Socony until
        1947)
    (3) Standard Oil of California
    (4) The Texas Company
    (5) Standard Oil Company (New Jersey)
    (6) Socony-Vacuum
    (7) Aramco (subsidiary of Socal and Texaco, and, after 1947,
        Jersey and Socony also)
2. Middle Eastern states (depositories of resources)
    (a) Saudi Arabia (dominated by 'Abd al-'Aziz)
    (b) Iraq (enemy of Saudi Arabia)
    (c) Levant states (Lebanon/Syria/Jordan)
    (d) Israel (opposed by Arab States)
3. Great Britain (a receding power)
    (a) The British government (the Foreign Office, Winston Churchill,
        and Lord Beaverbrook)
    (b) Anglo-Iranian and Royal Dutch-Shell
4. The Soviet Union (an emerging Cold War threat)

An intense effort was made to locate sufficient data to deal with the internal dynamics of each of these entities, with varying degrees of success. The reader may judge for himself, but I was most satisfied with the material that could be located on the Department of State, Harold Ickes, the British Foreign Office, Standard of California, Aramco, and 'Abd al-'Aziz. I was least satisfied with the material that could be acquired on the internal dynamics of The Texas Company and Socony-Vacuum. The rest of the material appeared to be adequate but not exceptional.

The next step was to study the primary and secondary data and to identify systematically, for each of the operative entities, its core belief system, its relative power vis-à-vis the others, its spatial context, its internal dynamics, and its specific objectives at each decision point. This was by far the most rewarding part of the process, because it produced a host of insights into relationships between pieces of data that might otherwise have gone unnoticed. The final step, of course, was to weave these insights into a narrative format reflecting evolutionary change over time. The intent was to have the narrative stand alone, free of methodological jargon, as a house stands alone when the scaffolding is removed.

To repeat what was said at the outset, the purpose of this brief account has been to place the work in intellectual context, clarify the perceptual screen through which the data has passed, and permit a critique of assumptions and method as well as conclusions.

APPENDIX B

TEXTS OF ANGLO-AMERICAN PETROLEUM
AGREEMENT, 1943, 1944, AND 1945

1. STATE DEPARTMENT DRAFT AGREEMENT WITH THE BRITISH ON PETROLEUM IN THE MIDDLE EAST (SEPTEMBER 16, 1943)

The Governments of the United States and of the United Kingdom, considering
  (a) The extreme importance of petroleum in wartime and in the post-war economy of all countries,
  (b) The fact that nationals of both countries hold—in part jointly—properties as well as rights to explore and develop petroleum-bearing areas in the Middle East (to be defined hereafter),
  (c) The present and post-war importance of assuring cooperation in the development of these holdings and the acquisition of new rights,
  (d) The necessity of assuring the proper conservation and development of petroleum resources for the maximum benefit of all countries, producers as well as consumers, have agreed as follows:

I.

The two Governments hereby establish and agree to maintain a Petroleum Board to be composed of ———members, ———members to be appointed by each Government.
  (a) The two Governments agree that they will invite governments of the producing areas covered by this agreement and other directly interested governments to participate in the activities of the Petroleum Board by means of suitable representation thereon.

II.

The two Governments agree that they will, through the Petroleum Board, concert their efforts to assure that the petroleum resources in the

Middle East (area to be defined) that may be held or acquired by either of them or by their nationals, will be developed as best to serve the following ends:

(a) That the oil resources in this area be developed in accordance with sound conservation practices.

(b) That the oil resources of the area are so developed as to promote the economic welfare of the people within whose territories they are located.

(c) That there be, in the event of an emergency, adequate available reserves for the military forces of the United Nations.

(d) That the two Governments may be assisted in ascertaining the practicability and the most expeditious means of arriving at an international agreement to deal with petroleum on a worldwide scale, the Petroleum Board will be expected to make suitable recommendations to the two Governments concerning a proposal of that nature to other governments.

III.

For the purposes set forth in Article II the two Governments agree to consult fully through the Petroleum Board regarding all plans and projects that they or their nationals may undertake for the operation or development of properties in regard to which they already possess rights in the Middle East, the relationships between the nationals of the two countries in such enterprises and undertakings, the exploration and the acquisition of new fields of operation, the construction of refinery facilities and pipelines, and the distribution of production.

IV.

It is agreed that the two Governments will keep the Petroleum Board fully advised regarding any matter discussed with any third government of concern to the Board and will afford the Board an opportunity to consider such matter.

V.

The Board shall make periodic reports to the two Governments.

In the event the United Nations establish a suitable economic council to take cognizance of such matters, the two Governments agree that such council will be given complete opportunity to investigate the activities of the Board and make recommendations with regard thereto.

VI.

The Petroleum Board shall make recommendations to the two Governments (a) concerning such executive or legislative action as it considers necessary to accomplish the purposes of this agreement, and (b) as regards such agreements with other governments as it considers necessary to serve the desired ends.

VII.

Both Governments undertake to use their best efforts to assure that oil produced in this area will be made freely available to all United Nations and other friendly countries on completely equal terms and conditions.

VIII.

Nothing in this Agreement shall be construed as affecting the sovereignty of the political entities of the area to which it refers.

SOURCE: Enclosure in letter, Wright to Stettinius, December 24, 1943, unmarked black binder, box 3, Records of the Petroleum Division, RG59.

2. FIRST ANGLO-AMERICAN PETROLEUM AGREEMENT (AUGUST 8, 1944)

Agreement on Petroleum between the Government of the United States of America and the Government of the United Kingdom of Great Britain and Northern Ireland.

INTRODUCTORY ARTICLE.

The Government of the United States of America and the Government of the United Kingdom of Great Britain and Northern Ireland, whose nationals hold, to a substantial extent jointly, rights to explore and develop petroleum resources in other countries, recognise:

1. That ample supplies of petroleum, available in international trade to meet increasing market demands, are essential for both the security and economic well-being of nations;
2. That for the foreseeable future the petroleum resources of the world are adequate to assure the availability of such supplies;
3. That such supplies should be derived from the various producing areas of the world with due consideration of such factors as available reserves, sound engineering practices, relevant economic factors,

and the interests of producing and consuming countries, and with a view to the full satisfaction of expanding demand;

4. That such supplies should be available in accordance with the principles of the Atlantic Charter and in order to serve the needs of collective security;

5. That the general adoption of these principles can best be promoted by international agreement among all countries interested in the petroleum trade whether as producers or consumers.

ARTICLE I.

The two Governments agree that the development of petroleum resources for international trade should be expanded in an orderly manner on a world-wide basis with due consideration of the factors set forth in paragraph 3 of the Introductory Article and within the framework of applicable laws or concession contracts. To this end, and as a preliminary measure to the calling of the international conference referred to in Article II below, the two Governments will so direct their efforts, with respect to petroleum resources in which rights are held or may be acquired by the nationals of either country:

1. That, subject always to considerations of military security and to the provisions of such arrangements for the preservation of peace and prevention of aggression as may be in force, adequate supplies of petroleum shall be available in international trade to the nationals of all peaceable countries at fair prices and on a non-discriminatory basis;

2. That the development of petroleum resources and the benefits received therefrom by the producing countries shall be such as to encourage the sound economic advancement of those countries;

3. That the development of these resources shall be conducted with a view to the availability of adequate supplies of petroleum to both countries as well as to all other peaceable countries, subject to the provisions of such collective security arrangements as may be established;

4. That, with respect to the acquisition of exploration and development rights in areas not now under concession, the principle of equal opportunity shall be respected by both Governments;

5. That the Government of each country and the nationals thereof shall respect all valid concession contracts and lawfully acquired rights, and shall make no effort unilaterally to interfere directly or indirectly with such contracts or rights;

6. That, subject always to the considerations mentioned in paragraph I of this Article, the exploration for and development of petroleum resources, the construction and operation of refineries and other facilities, and the distribution of petroleum shall not be hampered by restrictions imposed by either Government or its nationals, inconsistent with the purposes of this Agreement.

ARTICLE II.

The two Governments recognise that the principles declared in Article I hereof are of general applicability and merit adherence on the part of all countries interested in the international petroleum trade of the world.

Therefore, with a view to the wider adoption and effectuation of the principles embodied in this Agreement they agree that as soon as practicable they will propose to the Governments of other interested producing and consuming countries an International Petroleum Agreement which, *inter alia*, would establish a permanent International Petroleum Council composed of representatives of all signatory countries.

To this end the two Governments hereby pledge themselves to formulate plans for an international conference to consider the negotiation of such a multilateral Petroleum Agreement. They also pledge themselves to consult with other interested Governments with a view to taking whatever action is necessary to prepare for the proposed conference.

ARTICLE III.

There are, however, numerous problems of joint immediate interest to the two Governments, with respect to petroleum resources in which rights are held or may be acquired by their nationals, which must be discussed and resolved on a co-operative interim basis if the general petroleum supply situation is not to deteriorate.

With this end in view the two Governments hereby agree to establish an International Petroleum Commission to be composed of eight members, four members to be appointed immediately by each Government. This Commission, in furtherance of and in accordance with the principles stated in Article I hereof, shall consider problems of mutual interest to both Governments and their nationals, and, with a view to the equitable disposition of such problems, shall be charged with the following duties and responsibilities:

1. To prepare long-term estimates of world demand for petroleum, having due regard for the interests of consuming countries and expanding consumption requirements;

2. To suggest the manner in which, over the long term, this estimated demand may best be satisfied by production equitably distributed among the various producing countries in accordance with the criteria enumerated in paragraph 3 of the Introductory Article;
3. To recommend to both Governments broad policies for adoption by operating companies with a view to effectuating programmes suggested under the provisions of paragraph 2 of this Article;
4. To analyse such short-term problems of joint interest as may arise in connection with production, processing, transportation and distribution of petroleum on a world-wide basis, wherever the nationals of either country have a significant interest, and to recommend to both Governments such action as may appear appropriate;
5. To make regular reports to the two Governments concerning its activities;
6. To make, from time to time, such additional reports and recommendations to the two Governments as may be appropriate to carry out the purposes of this Agreement.

The Commission shall establish such organisation as is necessary to carry out its functions under this Agreement. The expenses of the Commission shall be shared equally by the two Governments.

ARTICLE IV.

To effectuate this Agreement the two Governments hereby grant reciprocal assurances:
1. That they will adhere to the principles set forth in Article I, paragraphs 1 to 6 inclusive;
2. That they will endeavour to obtain the collaboration of the Governments of other producing and consuming countries in the implementation of the principles set forth in Article I, and will consult, as appropriate, with such Governments in connexion with activities undertaken under Article III;
3. That upon approval of the recommendations of the Commission they will endeavour, in accordance with their respective constitutional procedures, to give effect to such approved recommendations;
4. That each Government will undertake to keep itself adequately informed of the current and prospective activities of its nationals with respect to the development, processing, transportation and distribution of petroleum;
5. That each Government will make available to the Commission such

information regarding the activities of its nationals as is necessary to the realisation of the purposes of this Agreement.

ARTICLE V.

The two Governments agree that in this Agreement:
1. The words "country" or "territories"
   (a) in relation to the Government of the United Kingdom of Great Britain and Northern Ireland, include, in addition to the United Kingdom, all British colonies, overseas territories, protectorates, protected States and all mandated territories administered by that Government; and
   (b) in relation to the Government of the United States of America, include, in addition to the United States, all territory under the jurisdiction of the United States;
2. The word "nationals" means
   (a) in relation to the Government of the United Kingdom of Great Britain and Northern Ireland, all British subjects and British protected persons belonging to the territories referred to in 1 (a) above and all companies incorporated under the laws of any of the above-mentioned territories, and also companies incorporated elsewhere in which the controlling interest is held by any of such nationals;
   (b) in relation to the Government of the United States of America, all nationals of the United States including companies incorporated under the laws of the territories referred to in 1 (b) above, and also companies incorporated elsewhere in which the controlling interest is held by any of such nationals;
3. The word "petroleum" means crude petroleum and its derivatives.

ARTICLE VI.

This Agreement shall enter into force upon a date to be agreed upon after each Government shall have notified the other of its readiness to bring the Agreement into force and shall continue in force until three months after notice of termination has been given by either Government or until it is superseded by the International Petroleum Agreement contemplated in Article II.

In witness whereof the undersigned, duly authorised thereto, have signed this Agreement.

Done in Washington, in duplicate, this eighth day of August, one thousand nine hundred and forty-four.

For the Government of the United States of America:

EDWARD R. STETTINIUS, Jr.,

Acting Secretary of State of the United States of America.

For the Government of the United Kingdom of Great Britain and Northern Ireland:

BEAVERBROOK, Lord Privy Seal.

MINUTES, PLENARY SESSION NO. V, ANGLO-AMERICAN CONVERSATIONS ON PETROLEUM, 3RD AUGUST, 1944, 4:30 P.M.

Conference Room, South Interior Building.

1.   In keeping with the earlier discussions on the subject of foreign exchange, the following conclusions were reached.

It was agreed by both Delegations that the terms of the Agreement provide for due consideration by the Commission of the foreign exchange position of each country.

The United Kingdom Delegation stated that during the post-war transitional period referred to in the draft Agreement for the International Monetary Fund, the United Kingdom might be obliged to take into account the exchange which it would lose or gain by the purchase or sale of petroleum in deciding the sources from which the petroleum it required should be drawn; that this situation would continue until sterling became freely convertible and all restrictions on payments and transfers for current international transactions have been removed.

The United States Delegation took note of this statement. They pointed out, however, that if the United Kingdom Government were to take unilateral action in this sense, the effect would be inconsistent with certain of the purposes of the Agreement.

It was agreed, therefore, that before any such action was taken the matter should be placed before the International Petroleum Commission, which, with due regard to the principles of the Agreement, should seek a solution acceptable to the two Governments; failing agreement on such a solution the Commission should consider the desirability of recommending to the two Governments a suspension of the Agreement in whole or in part. It was agreed that should the Commission be unable to make any recommendation acceptable to the two Governments, the Government presenting the problem to the Commission would exercise its right to give notice to terminate the Agreement in accordance with the provisions of Article VI. In order that such termination could take place within a reasonable length of time, it was agreed that term of notice in the Agreement should be shortened from six months to three months.

2. The two Delegations then agreed to submit the Agreement to their respective Governments with the recommendation that it should be entered into forthwith.

SOURCE: V42700,W12368/34/76,F0/371.

3. SECOND ANGLO-AMERICAN PETROLEUM AGREEMENT (SEPTEMBER 24, 1945)

PREAMBLE.

The Government of the United Kingdom of Great Britain and Northern Ireland and the Government of the United States of America, whose nationals hold, to a substantial extent jointly, rights to explore and develop petroleum resources in other countries, recognise:

1. That ample supplies of petroleum, available in international trade to meet increasing market demands, are essential for both the security and economic well-being of nations;
2. That for the foreseeable future the petroleum resources of the world are adequate to assure the availability of such supplies;
3. That the prosperity and security of all nations require the efficient and orderly development of the international petroleum trade;
4. That the orderly development of the international petroleum trade can best be promoted by international agreement among all countries interested in the petroleum trade, whether as producers or consumers.

The two Governments have therefore decided, as a preliminary measure to the calling of an international conference to consider the negotiation of a multilateral petroleum agreement, to conclude the following Agreement.

ARTICLE I.

The signatory Governments agree that the international petroleum trade in all its aspects should be conducted in an orderly manner on a worldwide basis with due regard to the considerations set forth in the Preamble, and within the framework of applicable laws and concession contracts. To this end and subject always to considerations of military security and to the provisions of such arrangements for the preservation of peace and prevention of aggression as may be in force, the signatory Governments affirm the following general principles with respect to the international petroleum trade:

(a) That adequate supplies of petroleum, which shall in this Agreement mean crude petroleum and its derivatives, should be accessible in international trade to the nationals of all countries on a competitive and non-discriminatory basis;

(b) That, in making supplies of petroleum thus accessible in international trade, the interests of producing countries should be safeguarded with a view to their economic advancement.

ARTICLE II.

In furtherance of the purposes of this Agreement, the signatory Governments will so direct their efforts:

(a) That all valid concession contracts and lawfully acquired rights shall be respected, and that there shall be no interference directly or indirectly with such contracts or rights;

(b) That with regard to the acquisition of exploration and development rights the principle of equal opportunity shall be respected;

(c) That the exploration for and development of petroleum resources, the construction and operation of refineries and other facilities, and the distribution of petroleum, shall not be hampered by restrictions inconsistent with the purposes of this Agreement.

ARTICLE III.

1. With a view to the wider adoption of the principles embodied in this Agreement, the signatory Governments agree that as soon as practicable they will propose to the Governments of all interested producing and consuming countries the negotiation of an International Petroleum Agreement, which *inter alia* would establish a permanent International Petroleum Council.

2. To this end the signatory Governments agree to formulate at an early date plans for an international conference to negotiate such a multilateral petroleum agreement. They will consult together and with other

interested governments with a view to taking whatever action is necessary to prepare for the proposed conference.

ARTICLE IV.

1. Numerous problems of joint immediate interest to the signatory Governments with respect to the international petroleum trade should be discussed and resolved on a co-operative interim basis if the general petroleum supply situation is not to deteriorate.

2. With this end in view, the signatory Governments agree to establish an International Petroleum Commission to be composed of six members, three members to be appointed immediately by each Government. To enable the Commission to maintain close contact with the operations of the petroleum industry, the signatory Governments will facilitate full and adequate consultation with their nationals engaged in the petroleum industry.

3. In furtherance of and in accordance with the purposes of this Agreement, the Commission shall consider problems of mutual interest to the signatory Governments and their nationals, and with a view to the equitable disposition of such problems it shall be charged with the following duties and responsibilities:

(a) To study the problems of the international petroleum trade caused by dislocations resulting from war;

(b) To study past and current trends in the international petroleum trade;

(c) To study the effects of changing technology upon the international petroleum trade;

(d) To prepare periodic estimates of world demands for petroleum and of the supplies available for meeting the demands, and to report as to means by which such demands and supplies may be correlated so as to further the efficient and orderly conduct of the international petroleum trade;

(e) To make such additional reports as may be appropriate for achieving the purposes of this Agreement and for the broader general understanding of the problems of the international petroleum trade.

4. The Commission shall have power to regulate its procedure and shall establish such organisation as may be necessary to carry out its functions under this Agreement. The expenses of the Commission shall be shared equally by the signatory Governments.

ARTICLE V.

The signatory Governments agree:
- (a) That they will seek to obtain the collaboration of the Governments of other producing and consuming countries for the realisation of the purposes of this Agreement, and to consult with such governments in connexion with activities of the Commission;
- (b) That they will assist in making available to the Commission such information as may be required for the discharge of its functions.

ARTICLE VI.

The signatory governments agree:
- (a) That the reports of the Commission shall be published unless in any particular case either Government decides otherwise;
- (b) That no provision in this Agreement shall be construed to require either Government to act upon any report or proposal made by the Commission, or to require the nationals of either Government to comply with any report or proposal made by the Commission, whether or not the report or proposal is approved by that Government.

ARTICLE VII.

The signatory Governments agree:
- (a) That the general purpose of this Agreement is to facilitate the orderly development of the international petroleum trade, and that no provision in this Agreement, with the exception of Article II, is to be construed as applying to the operation of the domestic petroleum industry within the country of either Government;
- (b) That nothing in this Agreement shall be construed as impairing or modifying any law or regulation, or the right to enact any law or regulation, relating to the importation of petroleum into the country of either Government;
- (c) That, for the purposes of this Article, the word "country" shall mean:
    - (i) in relation to the Government of the United Kingdom of Great Britain and Northern Ireland, the United Kingdom, those British colonies, overseas territories, protectorates,

protected states, and all mandated territories administered by that Government, and

(ii) in relation to the Government of the United States of America, the continental United States and all territory under the jurisdiction of the United States, lists of which, as of the date of this Agreement, have been exchanged.

ARTICLE VIII.

This Agreement shall enter into force upon a date to be agreed upon after each Government shall have notified the other of its readiness to bring the Agreement into force and shall continue in force until three months after notice of termination has been given by either Government or until it is superseded by the International Petroleum Agreement contemplated in Article III.

In witness whereof the undersigned, duly authorised thereto, have signed this Agreement.

Done in London, in duplicate, this twenty-fourth day of September, one thousand nine hundred and forty-five.

For the Government of the United Kingdom of Great Britain and Northern Ireland:

      (Signed)   EMANUEL SHINWELL

For the Government of the United States of America:

      (Signed)   HAROLD L. ICKES.

SOURCE: *British Command Paper No. 6683* (London: H. M. Stationery Office, 1945)

APPENDIX C

CRUDE POSITIONS OF JERSEY, SOCONY-VACUUM, SOCAL, AND THE TEXAS COMPANY, 1933-1950

The following tables are an attempt to place in perspective the relative crude positions of the four major partners in Aramco during the period under study. The statistics for the Standard Oil Company (New Jersey) are from the standard company history, and roughly comparable material was furnished to the author by Standard Oil of California. The data provided by the Mobil Oil Corporation were more sketchy, and Texaco declined to furnish anything beyond the limited material included in the 1953 history published by the company itself. As a result, comparable statistics are not available for all four companies.

The tables do, however, permit order-of-magnitude comparisons, and they do indicate that Jersey, and its junior partner Socony-Vacuum, were both crude-short during the entire period. Standard of California's gross production ran just under its refinery input until 1946, when Aramco's production converted its position into one of net crude surplus. All of this is at least consistent with the narrative in the body of the study. The data for The Texas Company is too sketchy to draw many conclusions, but its radically short crude position in the United States is consistent with its interest in joining Socal in the Casoc and Caltex ventures in 1936, and its emphasis on financial arrangements rather than market outlets for surplus crude in the 1940s.

TABLE C-1

STANDARD OIL COMPANY (NEW JERSEY) CRUDE POSITION, 1933-1950

(Daily average net crude oil production, purchases, sales, and net available to refineries, worldwide, in thousands of 42-gallon barrels.)

	Net Production of Affiliates[1]			Purchases from Nonaffiliates[2]			Sales to Nonaffiliates			Net Available to Affiliates' Refineries		
	United States	Foreign[3]	World[4]	United States	Foreign	World	United States	Foreign	World	United States	Foreign	World
1933	120.8	253.8	374.6	243.8	39.2	282.9	70.8	9.3	80.0	293.8	283.7	577.5
1934	131.4	324.0	455.5	222.8	45.7	268.5	82.0	10.7	92.7	272.2	359.1	631.3
1935	139.1	348.8	487.9	225.6	50.6	276.2	104.2	12.5	116.7	260.5	386.9	647.4
1936	155.3	367.6	523.0	270.1	51.3	321.4	137.9	11.5	149.4	287.6	407.5	695.0
1937	186.2	415.6	601.9	340.7	54.4	395.0	147.9	13.1	161.1	378.9	456.8	835.7
1938	166.1	395.4	561.5	319.9	52.5	372.3	126.4	17.4	143.6	359.6	430.6	790.1
1939	185.1	430.6	615.7	329.5	62.7	392.2	129.8	34.7	164.5	384.7	458.5	843.3
1940	206.1	385.9	592.0	346.2	59.0	405.2	139.9	34.3	174.2	412.4	410.4	822.8
1941	219.1	456.6	675.7	382.1	67.7	449.8	155.6	40.8	196.4	445.6	483.5	929.1
1942	216.8	258.7	475.5	404.6	58.1	462.7	168.5	19.5	188.0	452.9	297.3	750.2
1943	296.2	302.6	598.8	510.3	92.3	602.6	294.3	29.3	323.6	512.2	365.5	877.7
1944	366.8	422.6	789.4	674.7	117.0	791.7	443.7	56.4	500.1	597.8	483.2	1,081.0
1945	361.1	490.1	851.2	681.0	137.7	818.7	451.7	88.2	539.9	590.4	539.5	1,129.9
1946	366.7	563.7	930.4	627.8	150.1	777.9	382.5	88.5	471.0	404.9	832.4	1,237.3
1947	397.6	616.6	1,014.2	783.5	148.9	932.4	491.7	105.3	597.0	689.4	660.2	1,349.6
1948	427.9	801.0	1,228.9	908.2	130.6	1,038.8	604.1	120.8	724.9	732.0	710.8	1,442.8
1949	337.2	838.7	1,175.9	712.4	146.3	858.7	446.8	115.4	562.2	602.8	869.5	1,472.3
1950	342.5	946.6	1,289.1	737.6	167.3	904.9	411.9	120.7	532.6	667.4	1,014.9	1,682.3

SOURCE: Larson, Knowlton, and Popple, pp. 148, 474, 720-21.

1 Gross production minus royalty oil or oil belonging to other interests.
2 Includes royalty oil purchased as well as purchases from nonaffiliates.
3 Includes Jersey share in Aramco production beginning in 1947.
4 Because of rounding, world figures do not necessarily equal the exact sum of the two preceding figures.

TABLE C-2

SOCONY-VACUUM OIL COMPANY CRUDE POSITION, 1933-1950

(Daily average net crude oil production, and petroleum product sales, worldwide, in thousands of 42-gallon barrels.)

	Net Crude Oil Production[1]			Petroleum Product Sales		
	United States and Canada	Other Countries	World	United States and Canada	Other Countries	World
1933	89.7	N.A.	N.A.	N.A.	N.A.	N.A.
1934	96.1	16.9	113.0	N.A.	N.A.	N.A.
1935	103.8	27.9	131.7	N.A.	N.A.	N.A.
1936	113.1	31.0	144.1	N.A.	N.A.	N.A.
1937	139.7	32.6	172.3	N.A.	N.A.	N.A.
1938	131.3	31.5	162.8	256	95	382
1939	146.8	35.0	181.8	282	87	401
1940	151.3	37.1	188.4	303	72	406
1941	142.8	33.7	176.5	N.A.	N.A.	N.A.
1942	141.6	13.4	155.0	N.A.	N.A.	N.A.
1943	154.6	14.8	169.3	N.A.	N.A.	N.A.
1944	168.5	21.0	189.5	N.A.	N.A.	N.A.
1945	168.7	28.3	197.0	N.A.	N.A.	N.A.
1946	166.3	39.0	205.3	446	82	559
1947	167.5	67.9	235.4	495	99	634
1948	185.4	110.4	295.8	486	105	636
1949	168.4	140.6	309.0	468	124	645
1950	181.2	159.4	340.6	507	149	713

SOURCE: Letter to the author from C. R. Williams, Mobil Oil Corporation, March 3, 1978.

1. Excludes royalties and retained production payments owned or held by others.

TABLE C-3
STANDARD OIL COMPANY OF CALIFORNIA CRUDE POSITION, 1933-1950

(Daily average of gross liquids production, crude purchases and sales, and refinery input, worldwide, in thousands of 42-gallon barrels.)

	Gross Liquids Production[1]			Crude Purchases			Crude Sales			Refinery Input		
	United States	Foreign[2]	World	United States	Foreign	World	United States	Foreign	World	United States	Foreign	World
1933	99	0	99	39	0	39	18	7	25	110	0	110
1934	103	1	104	41	0	41	15	10	25	115	0	115
1935	122	3	125	42	0	42	18	21	39	123	0	123
1936	118	9	127	54	0	54	19	13	32	137	2	139
1937	132	11	143	62	0	62	19	16	35	147	6	153
1938	123	12	135	73	0	73	22	24	46	136	12	148
1939	105	16	121	86	0	86	25	16	41	139	16	155
1940	104	17	121	88	0	88	21	16	37	141	19	160
1941	114	15	129	90	0	90	14	7	21	157	18	175
1942	136	15	151	113	0	113	17	4	21	185	18	203
1943	164	16	180	119	0	119	21	3	24	226	18	244
1944	196	20	216	122	0	122	31	2	33	244	23	267
1945	228	40	268	111	0	111	47	3	50	256	41	297
1946	246	94	340	119	2	121	72	8	80	243	92	335
1947	276	88	364	141	3	144	88	8	96	271	100	371
1948	305	134	439	154	14	168	104	8	112	307	117	424
1949	296	164	460	144	19	163	111	11	122	301	121	422
1950	298	194	492	166	29	195	143	12	155	302	131	433

SOURCE: Comptroller's Department, Standard Oil Company of California; letter to the author from W. K. Morris, Standard Oil Company of California, February 6, 1978.

1. Crude plus butanes and propanes obtained as byproducts of natural gas production.   2. Includes Socal share in Aramco production.

TABLE C-4

THE TEXAS COMPANY CRUDE POSITION, 1933-1950

(Daily average of net crude oil and condensate production, worldwide, and
refinery runs to stills in the United States,
in thousands of 42-gallon barrels)

	Net Crude Oil and Condensate Production			Refinery Runs to Stills
	United States	Foreign	World-wide	United States Only
1933	84.5	0.5	85.0	179.3
1934	85.2	0.5	85.7	194.9
1935	103.1	0.2	103.3	209.0
1936	110.0	0	110.0	241.5
1937	135.7	0	135.7	264.2
1938	131.9	0	131.9	259 9
1939	173.8	0	173.8	275.3
1940	231.8	0	231.8	278.5
1941	202.0	0	202.0	315.5
1942	178.9	0	178.9	277.3
1943	196.9	0.1	197.0	310.0
1944	229.7	0.4	230 1	363.6
1945	235.3	1.5	236.8	354.6
1946	230.3	2.9	237.4	363.6
1947	245.9	7.1	253.0	381.2
1948	263.9	10.6	274.5	399.9
1949	243.4	20.4	263.8	413.8
1950	258.2	23.7	281.9	375.1

SOURCE: Adapted from Marquis James, *The Texaco Story* (n.p.: The Texas Company, 1953), pp. 108-109.

ESSAY ON SOURCES

THE PROCESS of collecting data for this study proved to be almost as interesting as the subject matter itself. A listing of the principal materials located is given below, but a comment on those items that proved to be of most value is in order.

On the government side the starting point was Record Group 59, the General Records of the Department of State in the National Archives in Washington, D.C. In addition to the voluminous and well-organized decimal files, a number of special lot files, especially the one for the Petroleum Division, proved to be invaluable. From there the search broadened to the records of other government agencies deposited in the National Archives, and the single most valuable find was a set of original records from the Petroleum Reserves Corporation, found buried in Record Group 234, the Records of the Reconstruction Finance Corporation. These official records were supplemented by a search of the personal papers of a number of key participants—chiefly at the Franklin D. Roosevelt Library in Hyde Park, New York, and the Harry S Truman Library in Independence, Missouri. The most valuable of this material proved to be the diary of Harold L. Ickes at the Library of Congress in Washington, D.C. Ickes dictated his peppery recollection of the week's events each weekend, and the diary makes fascinating reading in its own right. For the British side of the story, a considerable number of documents were found in British Foreign Office Political Correspondence, Class FO/371, in the Public Record Office in London.

The corporate side of the story proved to be considerably more challenging. As the story unraveled, it became clear that records and recollections of primary interest were those from the four parent companies, because this was where key decisions were made in the period under study. Until the late 1940s Aramco's corporate headquarters were in San Francisco, and they were for all practical purposes simply an adjunct of Socal's production department. The most interesting Aramco records would have been the personal files of James Terry Duce in Washington, but these had been destroyed when he retired. None of the four companies granted direct personal access to corporate archives, but a very large body of data proved to be available thanks to the subpoena power of the

United States Senate and the Federal Trade Commission, the willingness of a number of former company officials to share their recollections in personal interviews, and the cooperativeness of Professor Bennett H. Wall of Tulane University.

The approach taken was to collect documents and then to interview participants on what they recalled about the events recorded in those documents. For the early part of the period, the most valuable documents were those collected by the Senate Special Committee Investigating the National Defense Program and the Senate Special Committee Investigating Petroleum Resources just after World War II. For the later period, the most valuable were the documents (primarily from Socal) assembled by Senator Frank Church's Subcommittee on Multinational Corporations of the Committee on Foreign Relations in 1974. An additional source was the collection of corporate documents assembled by the Federal Trade Commission in the process of researching its report on *The International Petroleum Cartel* in the early 1950s. These were made available on a Freedom of Information Act request and proved to be a very extensive collection—chiefly photostats of documents from Exxon. In addition to all of this, Exxon agreed to let Professor Wall share with me whatever he could locate on the subject in connection with his research in process on a fourth volume of the history of Exxon. Professor Wall provided me with an extensive collection of excerpts from minutes of the Exxon executive committee and board of directors, and reported that almost all other Exxon records from that period had been destroyed. Curiously, this suggests that there may now be more data in the Federal Trade Commission collection of photostats than in the files of Exxon itself.

With all of this material in hand, the personal interviews proved to be most rewarding. The most valuable, of course, were with people who had been direct participants in the events in question. The single most valuable interview was with R. Gwin Follis of Socal, who had been intimately involved in almost the entire sequence of events. Particularly helpful for specific episodes were Augustus C. Long of Texaco; John C. Case of Mobil; Edward F. Johnson, David A. Shepard, Smith D. Turner, and Howard W. Page of Exxon; and Thomas C. Barger, Floyd W. Ohliger, Robert I. Brougham, George W. Ray, Jr., and Douglas Erskine of Aramco. The picture was rounded out through twenty-two additional interviews with former corporate and government officials familiar with, but less closely involved in, the events in question. Taken as a whole, this body of documents and interviews proved extremely useful in piecing together the corporate side of the story.

SELECTED BIBLIOGRAPHY

RECORDS AT THE UNITED STATES NATIONAL ARCHIVES,
WASHINGTON, D.C.

National Security Council Numbered Papers, 1947-1953 (Collection of declassified copies held by the Modern Military Branch).

Record Group 48, Records of the Department of the Interior.

Record Group 59, General Records of the Department of State.

Record Group 80, General Records of the Department of the Navy.

Record Group 84, Diplomatic Post Files, Dhahran.

Record Group 122, Records of the Federal Trade Commission.

Record Group 218, Records of the Joint Chiefs of Staff.

Record Group 234, Records of the Reconstruction Finance Corporation, 1932-1964 (Records Regarding Claim Against the Petroleum Reserves Corporation by the Arabian American Oil Company).

Record Group 250, Records of the Office of War Mobilization and Reconversion.

Record Group 253, Records of the Petroleum Administration for War.

Record Group 330, Records of the Office of the Secretary of Defense.

Record Group 353, Records of the State-War-Navy Coordinating Committee.

Record Group 428, General Records of the Navy Department, 1947–.

COLLECTIONS OF PERSONAL PAPERS

Dean Acheson Papers, Harry S. Truman Library, Independence, Missouri.

William L. Clayton Papers, Harry S. Truman Library, Independence, Missouri.

Herbert Feis Papers, Library of Congress, Washington, D.C.

James V. Forrestal Papers, Mudd Library, Princeton University, Princeton, New Jersey.

Harold L. Ickes Papers, Library of Congress, Washington, D.C.

Franklin D. Roosevelt Papers, Franklin D. Roosevelt Library, Hyde Park, New York.

Harry S. Truman Papers, Harry S. Truman Library, Independence, Missouri.

Fred M. Vinson Papers, University of Kentucky Library, Lexington, Kentucky.

MISCELLANEOUS ARCHIVAL COLLECTIONS

British Foreign Office Political Correspondence, Class FO/371, British Public Record Office, London, England

Records of the Exxon Corporation, New York, New York (excerpts from minutes of the executive committee and board of directors pertaining to Aramco and Middle Eastern oil, 1945-1950).

Records of the Federal Trade Commission, Washington, D.C. (corporate documents relating to the report on *The International Petroleum Cartel*).

Records of the Mobil Oil Corporation, New York, New York (production statistics and annual reports).

Records of the Office of the Chief of Naval Operations, Navy Historical Office, Navy Yard, Washington, D.C.

Records of the Standard Oil Company of California, San Francisco, California (production statistics and annual reports).

Records of Texaco, Inc., New York, New York (data provided to the author on the history of Aramco).

GOVERNMENT DOCUMENTS

Great Britain. *Papers by Command*, nos. 6551-6577, and 6659-6687. London: H. M. Stationery Office, 1944/1945.

Great Britain. *Parliamentary Debates*, 1946. London: H. M. Stationery Office, 1946.

Saudi Arabia. *Arbitration Between the Government of Saudi Arabia and Arabian American Oil Company*. Vol. 5. [First] *Memorial of Arabian American Oil Company*. Geneva: n.p., 1956. Copy in Law Library, University of California at Berkeley.

U.S. Congress. "An Act to Promote Export Trade, and for Other Purposes [Webb-Pomerene Act]." *United States Statutes at Large*. Vol. 40, chap. 50 (1918), pp. 516-18.

U.S. Congress. *Congressional Record*. 78th Congress, 2d sess. (1944) through 81st Congress, 1st sess. (1949).

U.S. Congress. Senate. Committee on Foreign Relations. *Anglo-American Oil Agreement*, Report to Accompany Executive H, 79th Cong., 1st sess. Executive Report no. 8, 80th Cong., 1st sess. Washington, D.C.: Government Printing Office, 1947.

U.S. Congress. Senate. Committee on Foreign Relations. *Petroleum Agreement with Great Britain and Northern Ireland, Hearings* (June 2-25, 1947). 80th Cong., 1st sess. Washington, D.C.: Government Printing Office, 1947.

U.S. Congress. Senate. Committee on Foreign Relations. Subcommittee on Multinational Corporations. *A Documentary History of the Petroleum Reserves Corporation, 1943-1944.* Committee Print, May, 1974. 93d Cong., 2d sess. Washington, D.C.: Government Printing Office, 1974.

U.S. Congress. Senate. Committee on Foreign Relations. Subcommittee on Multinational Corporations. *Multinational Corporations and United States Foreign Policy, Hearings*, pts. 4, 7, and 8 (January 28-March 28, 1974). 93d Cong., 2d sess. Washington, D.C.: Government Printing Office, 1974.

U.S. Congress. Senate. Committee on Foreign Relations. Subcommittee on Multinational Corporations. *Multinational Oil Corporations and U.S. Foreign Policy, Report.* Committee Print, January, 1975. 93d Cong., 2d sess. Washington, D.C.: Government Printing Office, 1975.

U.S. Congress. Senate. Committee on Interior and Insular Affairs. *Geopolitics of Energy.* By Melvin A. Conant and Fern R. Gold. Energy Publication 95-1. Committee Print, January, 1977. 95th Cong., 1st sess. Washington, D.C.: Government Printing Office, 1977.

U.S. Congress. Senate. Committee on the Judiciary and Committee on Interior and Insular Affairs. *The Emergency Oil Lift Program and Related Oil Problems, Joint Hearings* (February 27-March 22, 1957), pt. 2. 85th Cong., 1st sess. Washington: Government Printing Office, 1957.

U.S. Congress. Senate. *Petroleum Agreement with Great Britain and Northern Ireland.* Executive H, 79th Cong., 1st sess. Washington, D.C.: Government Printing Office, 1945.

U.S. Congress. Senate. Select Committee on Small Business. Subcommittee on Monopoly. *The International Petroleum Cartel.* Staff Report to the Federal Trade Commission [1952]. Reprint, April 1975. 94th Cong., 1st sess. Washington, D.C.: Government Printing Office, 1975.

U.S. Congress. Senate. Special Committee Investigating the National Defense Program. *Navy Purchases of Middle East Oil.* Report no. 440, pt. 5. 80th Cong., 2d sess. Washington, D.C.: Government Printing Office, 1948.

U.S. Congress. Senate. Special Committee Investigating the National Defense Program. *Petroleum Arrangements with Saudi Arabia, Hear-*

ings, pt. 41. 80th Cong., 1st sess. Washington, D.C.: Government Printing Office, 1948.

U.S. Congress. Senate. Special Committee Investigating the National Defense Program. *Report of Subcommittee Concerning Investigations Overseas, Section 1, Petroleum Matters*. Report no. 10, pt. 15. 78th Cong., 2d sess. Washington, D.C.: Government Printing Office, 1944.

U.S. Congress. Senate. Special Committee Investigating Petroleum Resources. *American Petroleum Interests in Foreign Countries, Hearings*. 79th Cong., 1st sess. Washington: Government Printing Office, 1946.

U.S. Congress. Senate. Special Committee Investigating Petroleum Resources. *Diplomatic Protection of American Petroleum Interests in Mesopotamia, Netherlands East Indies, and Mexico*. Study prepared by Henry S. Fraser. Senate Document no. 43. 79th Cong., 1st sess. Washington, D.C.: Government Printing Office, 1945.

U.S. Congress. Senate. Special Committee Investigating Petroleum Resources. *The Independent Petroleum Company, Hearings* (March 19-28, 1945). 79th Cong., 2d sess. Washington, D.C.: Government Printing Office, 1946.

U.S. Congress. Senate. Special Committee Investigating Petroleum Resources. *Intermediate Report*. Report no. 179. 79th Cong., 1st sess. Washington, D.C.: Government Printing Office, 1945.

U.S. Congress. Senate. Special Committee Investigating Petroleum Resources. *Petroleum Requirements-Postwar, Hearings* (October 3-4, 1945). 79th Cong., 1st sess. Washington, D.C.: Government Printing Office, 1945.

U.S. Department of State. *Foreign Relations of the United States*. Washington, D.C.: Government Printing Office, 1852-; exact title varies.

U.S. District Court of Southern New York, *File Civ. 39-779* (Suit of James A. Moffett, 1948).

INTERVIEWS WITH FORMER CORPORATE AND GOVERNMENT OFFICIALS

Arabian American Oil Company

Thomas C. Barger, in La Jolla, California, August 26, 1977.

Robert I. Brougham, by telephone to La Jolla, California, December 2, 1977.

Douglas Erskine, by telephone to Portola Valley, California, December 28, 1977.

William E. Mulligan, by telephone to New Boston, New Hampshire, December 16, 1977.

Floyd W. Ohliger, in Pineville, Pennsylvania, July 21, 1977.

George W. Ray, Jr., by telephone to East Thetford, Vermont, February 11, 1978.

George Rentz, in Washington, D.C., December 27, 1976.

Department of State

Emilio G. Collado (in the Department of State in 1945 and subsequently a vice-president of Exxon), by telephone to New York, New York, July 6, 1977.

Richard Funkhouser, by correspondence to Edinburgh, Scotland, September-November, 1977.

Raymond A. Hare, in Washington, D.C., July 15, 1977.

Parker T Hart, in Washington, D.C., July 15, 1977.

Loy W. Henderson, in Washington, D.C., July 18, 1977.

Harry N. Howard, in Washington, D.C., July 19, 1977.

George C. McGhee, in Washington, D.C., July 19, 1977.

Edward S. Mason, by correspondence to Cambridge, Massachusetts, August, 1977.

James S. Moore, Jr., in Washington, Kentucky, April 21, 1977.

Joseph Satterthwaite, in Washington, D.C., July 19, 1977.

Willard L. Thorpe, by correspondence to Amherst, Massachusetts, September, 1977.

Department of the Navy

Carl McGowan, by telephone to Washington, D.C., July 6, 1978.

Department of the Treasury

George M. Bennsky, by telephone to Washington, D.C., January 24, 1978.

Exxon Corporation

Edward F. Johnson, in Scarsdale, New York, July 12, 1977.

Howard W. Page, in New York, New York, May 2, 1977.

David A. Shepard, in New York, New York, May 3, 1977.

Smith D. Turner, in Scarsdale, New York, July 12, 1977.

Iraq Petroleum Company

Stephen H. Longrigg, in Guildford, Surrey, England, June 12, 1977.

Mobil Oil Corporation

John C. Case, in Keene Valley, New York, July 11, 1977.

Office of War Mobilization and Reconversion

Robert R. Nathan, by telephone to Washington, D.C., July 2, 1979.
Edward F. Prichard, Jr., by correspondence to Frankfort, Kentucky, May, 1978.

Standard Oil Company of California

George T. Ballou, in San Francisco, California, August 24, 1977.
R. Gwin Follis, in San Francisco, August 24, 1977.
Gaynor H. Langsdorf, by telephone to Hillsborough, California, December 3, 1977.
Theodore L. Lenzen, in Atherton, California, August 24, 1977.

Texaco, Inc.

Augustus C. Long, in New York, New York, August 3, 1977.

SECONDARY SOURCES

Abboushi, W[asif] F. *Political Systems of the Middle East in the 20th Century*. New York: Dodd, Mead, 1970.

Allison, Graham T. *Essence of Decision: Explaining the Cuban Missile Crisis*. Boston: Little, Brown, 1971.

Anderson, Irvine H. "Lend Lease for Saudi Arabia: A Comment on Alternative Conceptualizations." *Diplomatic History* 3, no. 4 (Fall, 1979), pp. 413-23.

Arabian Oil and It's Relation to World Oil Needs. Arabian American Oil Company, 1948.

Aramco Handbook: Oil and the Middle East. Rev. ed. Dhahran, Saudi Arabia: Arabian American Oil Company, 1968.

Armstrong, H[arold] C. *Lord of Arabia*. London: Arthur Barker, 1934.

Bandeau, John S. *The American Approach to the Arab World*. New York: Harper and Row, 1968.

Baram, Philip J. *The Department of State in the Middle East, 1919-1945*. Philadelphia: University of Pennsylvania Press, 1978.

Beaton, Kendall. *Enterprise in Oil: A History of Shell in the United States*. New York: Appleton-Century-Crofts, 1957.

Benoist-Méchin, Jacques. *Arabian Destiny*. Translated from the French by Denis Weaver. London: Elek Books, 1957.

Blair, John M. *The Control of Oil*. New York: Pantheon Books, 1976.

Bryson, Thomas A. *American Diplomatic Relations with the Middle East, 1784-1975: A Survey*. Metuchen, New Jersey: Scarecrow Press, 1977.

Carmical, J. H. *The Anglo-American Petroleum Pact: A Case Study in the Negotiation of Postwar Agreements*. New York: American Enterprise Association, 1945.

Cattan, Henry. *The Evaluation of Oil Concessions in the Middle East and North Africa*. Dobbs Ferry, New York: Oceana Publications, 1967.

Cheney, Michael S. *Big Oil Man From Arabia*. New York: Ballantine Books, 1958.

Childs, J. Rives. *Foreign Service Farewell: My Years of Service in the Near East*. Charlottesville, Virginia: University of Virginia Press, 1969.

The China White Paper, August 1949. 2 vols; Stanford: Stanford University Press, 1967.

Cook, M. A., ed. *Studies in the Economic History of the Middle East from the Rise of Islam to the Present Day*. London: Oxford University Press, 1970.

Cooper, Charles A., and Alexander, Sidney S., eds. *Economic and Population Growth in the Middle East*. New York: Elsevier, 1972.

Crozier, Michel. *The Bureaucratic Phenomenon*. Translated by the author. Chicago: University of Chicago Press, 1967.

Crum, Bartley C. *Behind the Silken Curtain: A Personal Account of Anglo-American Diplomacy in Palestine and the Middle East*. New York: Simon and Schuster, 1947.

Davies, Vincent. *Postwar Defense Policy and the U.S. Navy, 1943-1946*. Chapel Hill: University of North Carolina Press, 1966.

DeNovo, John A. *American Interests and Policies in the Middle East, 1900-1939*. Minneapolis: University of Minnesota Press, 1963.

———. "The Movement for an Aggressive American Oil Policy Abroad, 1918-1920." *American Historical Review* 61 (July, 1956), pp. 854-76.

Deutscher, Isaac. *Stalin: A Political Biography*. 2d ed. New York: Oxford University Press, 1966.

Dobney, Frederick J., ed. *Selected Papers of Will Clayton*. Baltimore: Johns Hopkins Press, 1971.

Donovan, Robert J. *Conflict and Crisis: The Presidency of Harry S. Truman, 1945-1948*. New York: Norton, 1977.

Ebenstein, William. *Modern Political Thought: The Great Issues*. 2d ed. New York: Holt, Rinehart, and Winston, 1960.

Eckes, Alfred E., Jr. *The United States and the Global Struggle for Minerals*. Austin, Texas: University of Texas Press, 1979.

Eddy, William A. *F.D.R. Meets Ibn Saud*. New York: American Friends of the Middle East, 1954.

Elwell-Sutton, Laurence P. *Persian Oil: A Study in Power Politics*. London: Lawrence and Wishart, 1955.

Engler, Robert. *The Politics of Oil: A Study of Private Power and Democratic Institutions*. New York: Macmillan, 1961.

Fairbank, John King. *The United States and China*. Rev. ed. New York: Viking Press, 1958.

Fanning, Leonard H. *Foreign Oil and the Free World*. New York: McGraw Hill, 1954.

Feis, Herbert. *The Birth of Israel: The Tousled Diplomatic Bed*. New York: Norton, 1969.

————. *Petroleum and American Foreign Policy*. Commodity Policy Studies no. 3. Stanford: Food Research Institute, Stanford University, 1944.

————. *Seen from E.A.: Three International Episodes*. New York: Alfred A. Knopf, 1947.

Finnie, David H. *Desert Enterprise: The Middle East Oil Industry in its Local Environment*. Cambridge: Harvard University Press, 1958.

Fisher, Sydney N. *The Middle East: A History*. 2d ed. New York: Alfred A. Knopf, 1969.

————. *Social Forces in the Middle East*. New York: Cornell University Press, 1955.

Ford, Alan W. *The Anglo-Iranian Oil Dispute of 1951-1952: A Study in the Role of Law in the Relations of States*. Berkeley: University of California Press, 1954.

Frey, John W., and Ide, H. Chandler. *A History of the Petroleum Administration for War, 1941-1945*. Washington, D.C.: Government Printing Office, 1946.

Gaddis, John Lewis. *The United States and the Origins of the Cold War, 1941-1947*. New York: Columbia University Press, 1972.

Gerretson, F. C. *History of the Royal Dutch*. 4 vols.; Leiden: E. J. Brill, 1953-57.

Gibb, George Sweet, and Knowlton, Evelyn H. *The Resurgent Years 1911-1927*. Vol. 2 in *History of Standard Oil Company (New Jersey)*. New York: Harper and Brothers, 1956.

Gibb, H.A.R. *Mohammedanism: An Historical Survey*. 2d ed. New York: Oxford University Press, 1935.

Graebner, Norman A. *Ideas and Diplomacy: Readings in the Intellectual*

Tradition of American Foreign Policy. New York: Oxford University Press, 1964.

Gulbenkian, Nubar. *Pantaraxia.* London: Hutchinson, 1965.

Halle, Louis J. *The Cold War as History.* New York: Harper and Row, 1967.

Halperin, Morton H. *Bureaucratic Politics and Foreign Policy,* with the assistance of Priscilla Clapp and Arnold Kantor. Washington, D.C.: Brookings Institution, 1974.

Hamilton, Charles W. *Americans and Oil in the Middle East.* Houston: Gulf Publishing Company, 1962.

Hartshorn, J. E. *Politics and World Oil Economics.* New York: Praeger, 1967.

Hegel, Georg Wilhelm Friedrich. *The Philosophy of History.* New York: Dover Publications, 1956.

Hidy, Ralph W., and Hidy, Muriel E. *Pioneering in Big Business, 1882-1911.* Vol. 1 in *History of Standard Oil Company (New Jersey).* New York: Harper and Brothers, 1955.

Hirst, David. *Oil and Public Opinion in the Middle East.* London: Faber and Faber, 1966.

Hofstadter, Richard. *The American Political Tradition and the Men Who Made It.* New York: Alfred A. Knopf, 1951.

Hopwood, Derek, ed. *The Arabian Peninsula: Society and Politics.* Totowa, New Jersey: Rowman and Littlefield, 1972.

Hull, Cordell. *The Memoirs of Cordell Hull.* 2 vols.; New York: MacMillan, 1948.

Huntington, Samuel P. *The Common Defense: Strategic Programs in National Politics.* New York: Columbia University Press, 1961.

Hurewitz, J[acob] C[oleman]. *Diplomacy in the Near and Middle East: A Documentary Record: 1914-1956.* Princeton, New Jersey: Van Nostrand, 1956.

————. *Middle East Dilemmas: The Background of United States Policy.* New York: Harper and Brothers, 1953.

————. *Middle East Politics: The Military Dimension.* New York: Praeger, 1969.

————. *The Struggle for Palestine.* New York: Norton, 1950.

Ickes, Harold L. *Fightin' Oil: Petroleum Administration for War.* New York: Alfred A. Knopf, 1943.

Ise, John. *The United States Oil Policy.* New Haven: Yale University Press, 1926.

Issawi, Charles, and Yeganeh, Mohammed. *The Economics of Middle Eastern Oil.* New York: Praeger, 1962.

Jacoby, Neil H. *Multinational Oil: A Study in Industrial Dynamics*. New York: Macmillan, 1974.

James, Marquis. *The Texaco Story: The First Fifty Years, 1902-1952*. The Texas Company, 1953.

Al-Jazairi, Mohammed Z. "Saudi Arabia: A Diplomatic History, 1924-1964." Ph.D. dissertation, University of Utah, 1971.

Katzenstein, Peter J., ed. *Between Power and Plenty: Foreign Economic Policies of Advanced Industrial States*. Madison: University of Wisconsin Press, 1978.

Kaufman, Burton I. *The Oil Cartel Case: A Documentary Study of Antitrust Activity in the Cold War Era*. Westport, Connecticut: Greenwood Press, 1978.

Kennan, George F. *American Diplomacy, 1900-1950*. Chicago: University of Chicago Press, 1951.

Kent, Marian. *Oil and Empire: British Policy and Mesopotamian Oil, 1900-1920*. London: Macmillan Press, 1976.

Kimcahe, Jon. *Seven Fallen Pillars: The Middle East, 1945-1952*. New York: Praeger, 1953.

Kirk, George E. *The Middle East in the War, Survey of International Affairs, 1939-1946, Volume II*. New York: Oxford University Press, 1952.

Klebanof, Shoshana. *Middle East Oil and U.S. Foreign Policy: With Special Reference to the Energy Crisis*. New York: Praeger, 1974.

Kolko, Gabriel. *The Politics of War: The World and United States Foreign Policy, 1943-1945*. New York: Random House, 1968.

——. *The Roots of American Foreign Policy: An Analysis of Power and Purpose*. Boston: Beacon Press, 1969.

Korb, Lawrence J. *The Joint Chiefs of Staff: The First Twenty-five Years*. Bloomington: Indiana University Press, 1976.

Krasner, Stephen D. *Defending the National Interest: Raw Materials Investments and U.S. Foreign Policy*. Princeton: Princeton University Press, 1978.

Krueger, Robert B. *The United States and International Oil: A Report for the Federal Energy Administration on U.S. Firms and Government Policy*. New York: Praeger, 1975.

Kuhn, Alfred. *The Logic of Social Systems: A Unified, Deductive, System-Based Approach to Social Science*. San Francisco: Jossey-Bass, 1974.

Kuhn, Thomas S. *The Structure of Scientific Revolutions*. 2d ed. Chicago: University of Chicago Press, 1970.

Kuniholm, Bruce R. *The Origins of the Cold War in the Near East:*

Great Power Conflict and Diplomacy in Iran, Turkey, and Greece. Princeton: Princeton University Press, 1980.

Larson, Henrietta M., Knowlton, Evelyn H., and Popple, Charles S. *New Horizons 1927-1950.* Vol. 3 in *History of Standard Oil Company (New Jersey).* New York: Harper and Row, 1974.

Lebkicher, Roy. *Aramco and World Oil.* New York: Russell F. Moore, 1952.

Lebkicher, Roy, Rentz, George, and Steineke, Max. *The Arabia of Ibn Saud.* New York: Russell F. Moore, 1952.

Lenczowski, George. *The Middle East in World Affairs.* 3rd ed. Ithaca: Cornell University press, 1962.

————. *Oil and State in the Middle East.* Ithaca: Cornell University Press, 1960.

Lenzen, Theodore L. "Inside International Oil," Unpublished typescript. Copy in library of the Council on World Affairs, San Francisco, Calif.

Levin, N. Gordon, Jr. *Woodrow Wilson and World Power: America's Response to War and Revolution.* New York: Oxford University Press, 1968.

Lijphart, Arend. "International Relations Theory," *International Relations Quarterly* 26, no. 1 (April, 1974), pp. 11-21.

————. "The Structure of the Theoretical Revolution in International Relations." *International Studies Quarterly* 18, no. 1 (March 1974), pp. 41-74.

Lipsky, George A. and others. *Saudi Arabia: Its People, Its Society, Its Culture.* New Haven: Human Relations Area Files Press, 1959.

Locke, John. *Locke on Politics, Religion, and Education.* Edited by Maurice Cranston. New York: Macmillan, 1965.

Longhurst, Henry. *Adventure in Oil: The Story of British Petroleum.* London: Sedgwick and Jackson, 1959.

Longrigg, Stephen Hemsley. *Oil in the Middle East: Its Discovery and Development.* 3rd ed. New York: Oxford University Press, 1968.

McClellan, Grant S., ed. *The Middle East in the Cold War.* New York: H. W. Wilson, 1956.

Magdoff, Harry. *The Age of Imperialism: The Economics of U.S. Foreign Policy.* New York: Monthly Review Press, 1969.

Mannheim, Karl. *Ideology and Utopia: An Introduction to the Sociology of Knowledge.* New York: Harcourt, Brace and World, 1936.

Mansfield, Peter, ed. *The Middle East: A Politicial and Economic Survey.* 4th ed. New York: Oxford University Press, 1973.

Maslow, A[braham] H. *Motivation and Personality*. New York: Harper and Row, 1954.

Mastny, Vojtech. *Russia's Road to the Cold War: Diplomacy, Warfare, and the Politics of Communism, 1941-1945*. New York: Columbia University Press, 1979.

Middle East Oil Developments: Summary. 2d ed. Arabian American Oil Company, 1948.

Mikdashi, Zuhayr M. *A Financial Analysis of Middle Eastern Oil Concessions, 1901-1965*. New York: Praeger, 1966.

Mikesell, Raymond F. *Foreign Investment in the Petroleum and Mineral Industries*. Baltimore: Johns Hopkins Press for Resources for the Future, 1971.

Mikesell, Raymond F., and Chenery, Hollis B. *Arabian Oil: America's Stake in the Middle East*. Chapel Hill, University of North Carolina Press, 1949.

Miller, Aaron D. *Search for Security: Saudi Arabian Oil and American Foreign Policy, 1939-1949*. Chapel Hill: University of North Carolina Press, 1980.

Millis, Walter, ed. *The Forrestal Diaries*. New York: Viking Press, 1951.

Monroe, Elizabeth. *Britain's Moment in the Middle East, 1914-1956*. Baltimore: Johns Hopkins Press, 1963.

Moore, Frederick L., Jr. "Origin of American Oil Concessions in Bahrein, Kuwait, and Saudi Arabia." Senior thesis, Princeton University, 1948.

Mosley, Leonard. *Power Play: Oil in the Middle East*. New York: Random House, 1973.

Mughraby, Muhamad A. *Permanent Sovereignty Over Oil Resources: A Study of Middle East Oil Concessions and Legal Change*. Beirut: Middle East Research and Publication Center, 1966.

Nash, Gerald D. *United States Oil Policy, 1890-1964: Business and Government in Twentieth Century America*. Pittsburgh: University of Pittsburgh Press, 1968.

Nietzsche, Friedrich. *The Will to Power*. Edited by Walter Kaufman. New York: Random House, 1967.

Notter, Harley. *Postwar Foreign Policy Preparation*. Department of State publication 3580, general foreign policy series 15. Washington: Government Printing Office, 1949.

Odell, Peter R. *An Economic Geography of Oil*. London: G. Bell and Sons, 1963.

―――. *Oil and World Power: A Geographical Interpretation*. New York: Taplinger, 1971.

Patai, Raphael. *The Arab Mind*. New York: Charles Scribner's Sons, 1973.

Peck, Malcolm C. "Saudi Arabia in United States Foreign Policy to 1958: A Study in the Sources and Determinants of American Policy." Ph.D. dissertation, Fletcher School of Law and Diplomacy, Tufts University, 1970.

Penrose, Edith T. *The Large International Firm in Developing Countries: The International Petroleum Industry*. London: Allen and Unwin, 1968.

Petroleum Facts and Figures, 1971 Edition. Washington, D.C.: American Petroleum Institute, 1971.

Philby, H[arry] St. John B[ridger]. *Arabian Days*. London: Robert Hale, 1948.

———. *Arabian Jubilee*. London: Robert Hale, 1952.

——— *Arabian Oil Ventures*. Washington, D.C.: Middle East Institute, 1964.

———. *Sa'udi Arabia*. London: Ernest Benn, 1955.

Podet, Allen H. "Anti-Zionism in a Key United States Diplomat: Loy Henderson at the End of World War II." *American Jewish Archives* 30, no. 2 (November, 1978), pp. 155-87.

Polk, William R. *The United States and the Arab World*. Rev. ed. Cambridge: Harvard University Press, 1969.

Report of Operations to the Saudi Arab Government—1949. Arabian American Oil Company.

Report of Operations to the Saudi Arab Government—1950. Arabian American Oil Company.

Rosenberg, David A. "The U.S. Navy and the Problem of Oil in a Future War: The Outline of a Strategic Dilemma, 1945-1950." *Naval War College Review* 29, no. 1 (Summer, 1976), pp. 53-64.

Sachar, Howard M. *Europe Leaves the Middle East, 1936-1954*. New York: Alfred A. Knopf, 1972.

Safran, Nadav. *The United States and Israel*. Cambridge: Harvard University Press, 1963.

Sampson, Anthony. *The Seven Sisters: The Great Oil Companies and the World They Made*. New York: Viking Press, 1975.

Sanger, Richard H. *The Arabian Peninsula*. Ithaca: Cornell University Press, 1954.

Sayegh, Kamal S. *Oil and Arab Regional Development*. New York: Praeger, 1968.

Schectman, Joseph B. *The United States and the Jewish State Movement: The Crucial Decade, 1939-1949*. New York: Herzl Press, 1966.

Schilling, Warner R., Hammond, Paul Y., and Snyder, Glenn H. *Strategy, Politics, and Defense Budgets*. New York: Columbia University Press, 1962.

Seale, Patrick. *The Struggle for Syria: A Study of Post-War Arab Politics, 1945-1958*. London: Oxford University Press, 1965.

Sherry, Michael S. *Preparing for the Next War: American Plans for Postwar Defense, 1941-45*. New Haven: Yale University Press, 1977.

Shwadran, Benjamin. *The Middle East, Oil, and the Great Powers*. 3rd rev. ed. New York: Halsted Press, 1973.

Smith, Adam. *An Inquiry Into the Nature and Causes of the Wealth of Nations*. Edited by Maurice Cranston. New York: Macmillan, 1965.

Snetsinger, John. *Truman, The Jewish Vote, and the Creation of Israel*. Stanford: Hoover Institution Press, 1974.

Spinoza, Baruch. *The Ethics of Spinoza: The Road to Inner Freedom*. Edited by Dagobert D. Runes. Secaucus, New Jersey: Citadel Press, 1976.

Stanley, Timothy W. *American Defense and National Security*. Washington, D.C.: Public Affairs Press, 1956.

Stegner, Wallace [E.]. *Discovery: The Search for Arabian Oil*. Beirut: Middle East Export Press, 1971.

Stevens, Richard P. *American Zionism and U.S. Foreign Policy, 1942-1947*. New York: Pageant Press, 1962.

Stocking, George W. *Middle East Oil: A Study in Political and Economic Controversy*. Nashville: Vanderbilt University Press, 1970.

Stoff, Michael B. *Oil, War, and American Security: The Search for a National Oil Policy, 1941-1947*. New Haven: Yale University Press, 1980.

Teilhard de Chardin, Pierre. *The Phenomenon of Man*. New York: Harper and Brothers, 1959.

Truman, Harry S. *Memoirs*. 2 vols. New York: Doubleday, 1956.

Turner, Louis. *Oil Companies in the International System*. London: George Allen and Unwin for the Royal Institute of International Affairs, 1978.

Twitchell, Karl S. *Saudi Arabia: With an Account of the Development of Its Natural Resources*. 2d ed. Princeton: Princeton University Press, 1953.

Van der Meulen, D. *The Wells of Ibn Saud*. London: John Murray, 1957.

Vernon, Raymond, ed. *The Oil Crisis*. New York: Norton, 1976.

Vicker, Ray. *The Kingdom of Oil: The Middle East: Its People and Its Power*. New York: Charles Scribner's Sons, 1974.

Walt, Joseph W. "Saudi Arabia and the Americans, 1928-1951." Ph.D. dissertation, Northwestern University, 1960.

Walz, Kenneth N. "Man, the State, and the State System in Theories of the Causes of War." Ph.D. dissertation, Columbia University, 1954.

Ward, Thomas E. *Negotiations for Oil Concessions in Bahrain, El Hasa (Saudi Arabia), the Neutral Zone, Qatar, and Kuwait.* New York: privately printed, 1965.

White, David H. *Proceedings of the Conference on War and Diplomacy, 1976.* Charleston, South Carolina: The Citadel, 1976.

White, Gerald T. *Formative Years in the West: A History of the Standard Oil Company of California and Predecessors Through 1919.* New York: Appleton-Century-Crofts, 1962.

Wilkins, Mira. *The Maturing of Multinational Enterprise: American Business Abroad from 1914 to 1970.* Cambridge: Harvard University Press, 1974.

Williams, William Appleman. *The Tragedy of American Diplomacy.* Rev. and enld. ed. New York: Dell, 1962.

Williamson, Harold F., et al. *The American Petroleum Industry.* 2 vols. Evanston: Northwestern University Press, 1959-63.

Wilmington, Martin W. *The Middle East Supply Center.* Edited by Laurence Evans. Albany: State University of New York Press, 1971.

Wilson, Joan Hoff. *American Business and Foreign Policy, 1920-1933.* Lexington: University of Kentucky Press, 1971.

Yergin, Daniel. *Shattered Peace: The Origins of the Cold War and the National Security State.* Boston: Houghton Mifflin, 1978.

INDEX

Library of Congress Cataloging in Publication Data

Anderson, Irvine H., 1928-
    Aramco, the United States, and Saudi Arabia.

    Bibliography: p.
    Includes index.
    1. Arabian American Oil Company.   2. Petroleum
industry and trade—Saudi Arabia.   3. Petroleum
industry and trade—United States.   4   United States
—Foreign economic relations—Saudi Arabia.
    5.   United States—Foreign relations—Saudi Arabia.
    I. Title.
HD9576.S35A72   338 7′6223382′09538   80-8535
ISBN 0-691-04679-4   AACR2

Milton Keynes UK
Ingram Content Group UK Ltd.
UKHW022128190524
442884UK00008B/466